AF768443

ALLE · ZEIT · WACH · 1842

Progress in

Sensory Physiology 2

Editors:
H. Autrum D. Ottoson E. Perl R. F. Schmidt

Editor-in-Chief: D. Ottoson

With Contributions by
G. A. Manley R. Necker C. A. Smith

With 103 Figures

Springer-Verlag
Berlin Heidelberg New York 1981

Editor-in-Chief:

Professor Dr. David Ottoson
Karolinska Institutet, Fysiologiska Institutionen II
Solnavägen 1, S-10401 Stockholm 60

Editors:

Professor Dr. Hansjochem Autrum
Zoologisches Institut der Universität München
Luisenstrasse 14, D-8000 München 2

Professor Dr. Edward Perl
Department of Physiology
University of North Carolina at Chapel Hill
Chapel Hill, NC 27514 (USA)

Professor Dr. Robert F. Schmidt
Physiologisches Institut der Universität
Olshausenstrasse 40 – 60, D-2300 Kiel

ISBN-13: 978-3-642-68171-4 e-ISBN-13: 978-3-642-68169-1
DOI: 10.1007/978-3-642-68169-1

This work is subject to copyright. All rights are reserved, whether the whole or part of the material is concerned, specifically those of translation, reprinting, re-use of illustrations, broadcasting, reproduction by photocopying machine or similar means, and storage in data banks. Under § 54 of the German Copyright Law, where copies are made for other than private use, a fee is payable to "Verwertungsgesellschaft Wort", Munich.

© by Springer-Verlag Berlin Heidelberg 1981

Softcover reprint of the hardcover 1st edition 1981

The use of registered names, trademarks, etc. in this publication does not imply, even in the absence of a specific statement, that such names are exempt from the relevant protective laws and regulations and therefore free for general use.

2121/3140-543210

Contents

Thermoreception and Temperature Regulation in Homeothermic Vertebrates

R. Necker

Institut für Tierphysiologie, Ruhr-Universität Bochum,
Postfach 102148, D-4630 Bochum 1

Temperature regulation in homeothermic vertebrates is based on sensory mechanisms and sensory neurons that detect temperature or temperature changes and on neuronal processing of temperature information in the central nervous system (CNS), which finally results in appropriate responses of the organism to keep the body temperature constant. This article concentrates on neuronal processes in the thermoregulatory system of birds and mammals; experiments in man have been considered only occasionally. Since there are several recent reviews (Hellon 1972b; Bligh 1973; Hensel 1973a, b, 1974; Simon 1974; Cabanac 1975) primary attention has been given to reports in the literature since 1973. Because temperature regulation in birds (Dawson and Hudson 1970; Richards 1975) has been shown to function essentially the same as that in mammals, both classes of homeothermic vertebrates will be treated here together and not separately.
Although several topics which are also clearly important from the neurophysiological point of view have been omitted, very recent reviews on these aspects of temperature regulation have been included. This article concentrates on autonomic thermoregulatory responses. Behavioral temperature regulation, which is not discussed here, has been shown to follow principally similar rules as autonomic regulation (Corbit 1970; Bligh 1973; Hensel 1973a; Cabanac 1979). As to putative neurotransmitter substances and pharmacological investigations there is a vast literature which will not be considered here (Hellon 1974; Bligh 1979; Clark 1979). Diurnal and seasonal variations in temperature regulation (sleep, daily torpor, hibernation) have received increasing interest. The neuronal aspects of this interesting topic have been reviewed recently (Heller and Glotzbach 1977; Heller et al. 1978; Heller 1979). Experiments with short-term deviations in body temperature (fever, exercise) have also been left out as have those investigating long-term adaptations (acclimation, acclimatization).
In temperature regulation it is very tempting to build models. Neuronal models have been proposed which are highly variable and which can be adapted easily to new results (Hammel 1972; Bligh 1972, 1979). Many variations have been published in the literature, but all these models usually explain only some results and

obviously are over simplifications of the complicated neuronal network involved in temperature regulation. As will be shown in this article we have gained some insight into this network, but we are far from understanding the whole system. Because models usually do not help to solve this problem, this article concentrates on describing results and avoids models.

1 Cutaneous Thermosensitivity

1.1 Cutaneous Thermoreceptors

Cutaneous thermoreceptors have been studied since the early 1950s and have been reviewed recently (Hensel 1973b, 1974, 1976; Duclaux 1977). In this chapter recent investigations and properties which may be important for temperature regulation are considered.

1.1.1 General Characteristics

Cutaneous thermoreceptors can be divided into two groups: cold receptors and warm receptors. According to Hensel (1973b) cutaneous thermoreceptors are characterized by the following properties: (1) there ist a spontaneous discharge over a wide temperature range, the firing rate under static conditions depending on skin temperature; (2) there ist a phasic response during dynamic thermal stimulation which is opposite in cold and warm receptors, the first being excited by cooling and the latter being inhibited, and vice versa, during warming; and (3) there is no response to mechanical stimulation of moderate or even strong intensity.

1.1.2 Structure

Cold Receptors. So far only the nerve terminals of cold receptors in the cat's nose have been identified (Hensel et al. 1974). Small myelinated fibers divide into a number of unmyelinated endings which penetrate the basal lamina of the epithelium and invaginate the cytoplasm of the basal epithelial cells. The endings contain numerous mitochondria and a receptor matrix.

Warm Receptors. Warm receptors of mammals are assumed to be located in deeper layers of the skin (Hensel 1973b), but there is as yet no successful identification of the nerve terminals. Facial warm receptors were identified in the pit organ of pit vipers (Terashima et al. 1970) and in the scales of the upper jaw of the boa constrictor (von Düring 1974). Myelinated fibers divide into a conglomeration of unmyelinated nerve branchlets which contain a high concentration of mitochondria. These receptors penetrate the epithelium and have myelinated axons and in this way differ from mammalian warm receptors.

1.1.3 Generator Processes

Since no one has ever succeeded in recording receptor potentials from the sensory region of thermoreceptors there is only indirect evidence for transducer mechanisms underlying thermoreceptor function. Sand (1938) proposed a model of cold sensitivity of the ampullae of Lorenzini which involved an excitatory and an inhibitory process. This model was adopted for cold receptor function (Hensel 1973b) and was recently extended by a depletion factor which, when described mathematically, closely fits the dynamic response of cold receptors under various conditions (Kenshalo et al. 1976). Physiological mechanisms underlying both excitatory and inhibitory processes have been studied, and those which deal with temperature-dependent membrane properties will be discussed here. One has to keep in mind, however, that this is only indirect evidence of thermoreceptor transducer processes.

Based on experiments on the temperature dependency of excitable membranes in *Aplysia* neurons (Carpenter 1970), the effect of ouabain and K^+-free solution on cold-sensitive afferents in the rat was studied (Pierau et al. 1974, 1975, 1980). Ouabain blocks the electrogenic Na^+ pump which itself is very sensitive to temperature changes (active transport) and which hyperpolarizes the excitable membrane. Perfusion of an isolated nerve-skin preparation (scrotal nerve of rats) with ouabain was followed by an excitation of cold receptors comparable to cooling. This effect was seen only at high skin temperatures where the pump is active. In this way cooling inhibits the Na^+ pump which ist followed by a depolarization resulting in an excitation (or the other way round: warming results in an inhibition). This hypothesis is supported by another experiment. The pump does not function in a K^+-free extracellular medium. When perfusing with K^+-free solution, again an excitation of the cold afferents occurred (Pierau et al. 1975). Similar results were found in frog cold afferents (Spray 1974b). The important role of metabolic energy (active transport) in cold receptor activity is additionally supported by experiments which show that an intact blood or oxygen supply is necessary for cold receptor function (Iggo and Paintal 1977).

As found in molluscan neurons (Gorman and Marmor 1970) there is another important temperature-dependent mechanism which influences membrane polarization: the permeability of Na^+ increases more than that of K^+ with increasing temperature (the ratio PNa^+/PK^+ shows a positive correlation with temperature). This results in an increasing depolarization with increasing temperature and counteracts the hyperpolarization of the Na^+ pump. This mechanism should dominate at low temperatures where the pump is nearly inactive and thus would explain the positive slope of static curves in cold receptors at low temperatures. No comparable investigations of warm receptors have been made, but one might speculate that the increase in the PNa^+/PK^+ ratio with increasing temperature may dominate in these endings also at high temperatures. But this is highly speculative at the moment.

Effects of Ca^{++}. Intravasal injection of Ca^{++} is followed by an excitation of warm receptors and an inhibition of cold receptors (Hensel and Schäfer 1974). In addition, it has been shown that Ca^{++} has an effect on the bursting pattern of cold receptors in that bursting is reduced at high Ca^{++} levels and enhanced after

Ca^{++} reduction by EDTA (Pierau and Wurster 1975; Schäfer et al. 1979; Pierau et al. 1980). No unequivocal explanation of all these results can be given at the moment. Nevertheless the Ca^{++} effect suggests that there might be a common basis for cold and warm receptor function. This is supported by the observation that the thermosensitivity of the ampullae of Lorenzini can change from cold sensitive to warm sensitive (Hensel and Nier 1976).

1.1.4 Receptive Field and Size of Afferent Fibers

In nonprimate mammals receptive fields are generally spotlike (<1 mm). In primates receptive fields of cold receptors can consist of several spots or small areas (Duclaux and Kenshalo 1973; Kenshalo and Duclaux 1977). As shown by Duclaux and Kenshalo (1973), there seems to be no spatial summation when stimulating more and more spots of the receptive field. The maximum response of the whole field is that of the most sensitive spot. In the bills of ducks the receptive fields of cold afferents had a mean size of 12 mm^2 (Leitner and Roumy 1974).
As shown by the conduction velocity all warm fibers belong to the C-fiber group (conduction velocity usually below 1 m/s). This was found in human warm fibers, as well (Konietzny and Hensel 1975). Part of the cold fibers belong to the C-fiber group, but part belong to the thinly myelinated Aδ-fiber group with conduction velocities up to about 40 m/s (Darian-Smith et al. 1973; Dykes 1975). The concept that cold fibers conduct at a higher velocity than warm fibers is supported by nerve block experiments in conscious man (Fruhstorfer et al. 1974; Fruhstorfer 1976) which show that warm sensation disappears before cold sensation (thin fibers being more sensitive to nerve block than thicker ones). Norrsell and Ullman (1978) calculated a conduction velocity below 1 m/s for warm receptors from the reaction time in man in psychophysical experiments. Conduction velocity of bird thermoafferent fibers has not been assessed as yet.

1.1.5 Dynamic Responses

Responses to temperature changes (dynamic responses) are generally phasic, both in cold and in warm receptors, i.e., there is a transient change in firing rate during the dynamic phase of the stimulus and a subsequent adaptation to a new level. Phasic responses depend on the stimulus magnitude (intensity) and on the rate of temperature change. In addition, the phasic response depends on the temperature range of the stimulus, and it has been shown in many investigations that phasic sensitivity is highest in the temperature range of the maximum of the static activity of thermoreceptors (Hensel 1973b).
One method of quantifying the excitatory phasic response consists in assessing the peak frequency. At a constant rate of temperature change there is a linear positive correlation between stimulus magnitude and peak frequency, usually only with small intensities, and a leveling off with higher intensities (see. Fig. 1). This was found both in cold receptors and in warm receptors (Hensel 1973b; Kenshalo and Duclaux 1977; Molinari and Kenshalo 1977; Duclaux and Kenshalo 1980; Konietzny and Hensel 1977; Hellon et al. 1975).

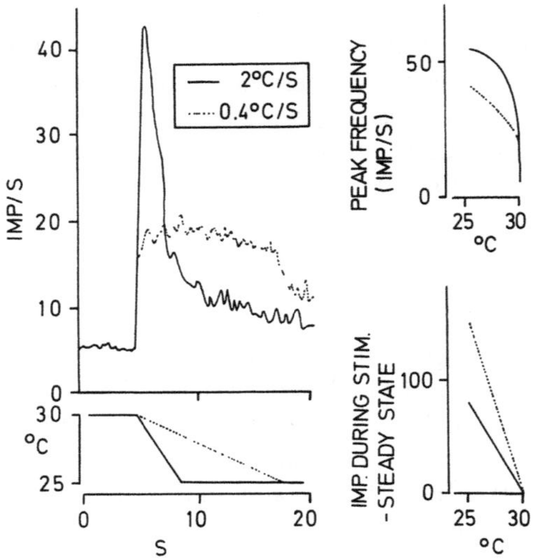

Fig. 1. Dynamic response of cold receptors. *Left side,* time course of the response (impulse/s) to a change of skin temperature from 30° to 25 °C with rates of temperature change of 2 °C/s and 0.4 °C/s. *Right side, upper diagram,* peak frequency of the dynamic response at increasing steps of cooling from 30 °C at two rates of temperature change. *Right side, lower diagram,* impulses during the temperature change minus steady state activity at the new level in dependence on increasing steps of cooling at two rates of temperature change; note the steeper increase of the response at 0.4 °C/s! Redrawn from Kenshalo and Duclaux 1977 (with permission)

Since the peak frequency does not represent the whole phasic response, another method of quantifying this response is to calculate a mean or cumulative response for some period after the beginning of the stimulus. In this case there is – at constant rates of temperature change – a linear correlation over a wider range of stimulus intensities than with the peak frequency, again both in cold and in warm receptors (Darian-Smith et al. 1973, 1979; Dykes 1975; Dubner et al. 1975).

Another method of quantifying the phasic response consists in assessing the total number of impulses after stimulation minus steady state activity at the new level (Kenshalo and Duclaux 1977; Duclaux and Kenshalo 1980). In this case the correlation was again linear over a rather wide temperature range (Fig. 1), comparable to the cumulative response described previously.

Only a few studies have investigated the dependence of the phasic response on the rate of temperature change. In monkeys there was a linear correlation only at low rates of temperature changes (<0.1 °C/s) and a plateau at higher rates (Molinari and Kenshalo 1977). When the total number of impulses minus steady state activity was assessed, a larger response at a lower rate (0.4 °C/s) than at a higher rate (2 °C/s) of temperature change (see Fig. 1) was found, both in cold receptors (Kenshalo and Duclaux 1977) and in warm receptors (Duclaux and Kenshalo 1980).

The inhibitory response to warming (cold receptors) or cooling (warm receptors) depends on the stimulus intensity in a similar manner as the excitatory response, but in general a plateau is reached at much smaller intensities (Darian-Smith et al. 1973; Dykes 1975; Kenshalo and Duclaux 1977; Duclaux and Kenshalo 1980). As to the dependence of the inhibitory response on the rate of temperature change no difference was found between 0.4 °C/s and 2 °C/s in cold receptors (Kenshalo and Duclaux 1977).

Altogether there seems to be a linear correlation between stimulus magnitude and phasic response, although peak frequency usually plateaus at large magnitudes (especially outside the range of maximal phasic sensitivity). With regard to the rate of temperature change the sensitivity of the phasic response seems to be best at low rates. These characteristics might be of importance for temperature regulation since under natural conditions homeothermic animals are usually subject to gradual skin temperature changes, especially when insulation (hair, feathers) reduces both the magnitude and rate of temperature changes. Because of the high sensitivity of the phasic response of thermoreceptors to small magnitudes and low rates of temperature changes this response may well contribute to temperature regulation.

1.1.6 Static Activity

Cold Receptors. The temperature range of static activity varies considerably for different cold fibers, the extremes being 5 °C and 43 °C (Hensel 1973b). In the golden hamster, a hibernator, the activity range extends to −5 °C (Raths and Hensel 1967). At the upper end of the activity range a total cessation of activity is not often found, but temperature sensitivity, instead, stops and in some receptors a "paradoxical" response, i.e., an increase in firing rate at noxious temperature levels, is observed (Dodt 1952; Kenshalo and Duclaux 1977; Long 1977).

The static curves of cold receptors usually reach a maximum at low skin temperatures (Fig.2). The temperature range where a negative correlation between skin temperature and firing rate exists extends from some upper temperature limit where, with increasing temperature, no further change in firing rate occurs up to the temperature of maximal static activity. The upper temperature limit differs from cold receptor to cold receptor: The range generally extends from about 35 ° to about 42 °C (Iggo 1969; Hensel and Wurster 1970; Kenshalo and Duclaux

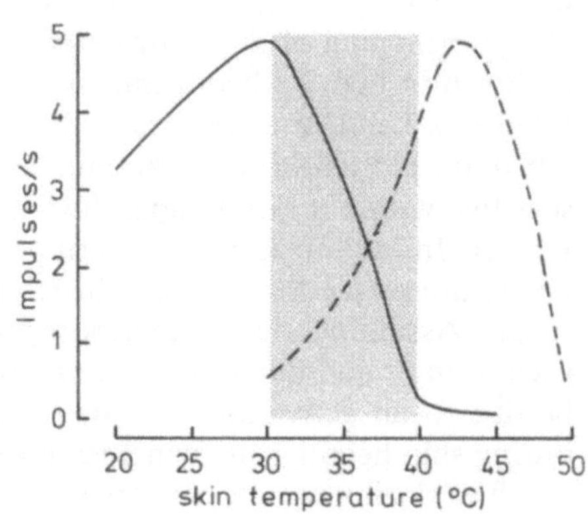

Fig. 2. Mean static curves of cold receptors and warm receptors *(dashed line)* in the skin of monkeys. The *stippled area* indicates the range of skin temperatures usually found in investigations on temperature regulation. Redrawn from Kenshalo and Duclaux 1977, and Duclaux and Kenshalo 1980 (with permission)

1977). From this it becomes evident that below about 42 °C more and more cold receptors are activated.
The range of maximal activity again varies considerably and extends from about 20 °C to about 30 °C (Iggo 1969; Hensel and Wurster 1970; Kenshalo and Duclaux 1977). Similar results were found in birds (Necker 1972; Gregory 1973) and in humans (Hensel and Boman 1960).
A characteristic feature of many cold receptors is a *burst discharge* at low skin temperature. Whereas the mean frequency shows a maximum near 30 °C the impulses per burst increase with decreasing temperature, even with temperatures below 20 °C (Iggo 1969; Dykes 1975; Hellon et al. 1975). This results in a prolongation of the burst duration since the intraburst interval does not change very much (Iggo 1969; Dykes 1975; Bade et al. 1979). The number of impulses per burst could thus signal a decrease in skin temperature at values at which the mean frequency does not change or decreases. Interestingly, the burst discharge was not found in second-order neurons in the CNS where static curves very similar to the mean frequency curve of peripheral receptors were found (Dostrovsky and Hellon 1978; Dickenson et al. 1979).

Warm Receptors. In the nasal area of the cat warm receptors become active between about 30° and 40 °C (Hensel 1973b), reach a maximum between 44° and 47 °C, and stop activity between about 45° and 48 °C, i.e., the range between maximum and cessation of activity at high temperatures is rather small. Scrotal warm receptors become active between about 30° and 38 °C, but their maxima are between about 40° and 44 °C (Hellon et al. 1975). Similar warm receptors were found in the monkey's hand and foot skin (Hensel and Iggo 1971; Duclaux and Kenshalo 1980; Darian-Smith et al. 1979). In addition to this predominantly found type, a second type was found in monkeys with a similar range of start of activity at low temperatures, but with maximal activity above 45 °C. These warm receptors differ from heat nociceptors mainly in that they show high sensitivity at skin temperatures below 40 °C (LaMotte and Campbell 1978). Warm fibers with a maximum at high skin temperatures (>45 °C) were found in humans (Konietzny and Hensel 1975) and in pigeons (Necker 1972). Interestingly, warm fibers in pigeons showed a burst discharge at low temperatures (Necker 1972, 1973), whereas in mammals this was observed at high skin temperatures (Hensel 1973b).

Comparison of Static Curves of Warm and Cold Receptors. There is a considerable overlapping of the static curves of warm receptors and cold receptors. This is shown in Fig. 2 where mean static curves of cold and warm receptors in the monkey's hand or foot (Kenshalo and Duclaux 1977; Duclaux and Kenshalo 1980) are plotted in one diagram. Both cold receptors and warm receptors are sensitive within a skin temperature range of about 30° – 40 °C (marked area in Fig. 2). In investigations of thermoregulatory responses to changes in ambient temperature (see Sect. 1.3), skin temperature variations are usually within this range. Assuming that a thermoregulatory response can be due to either an excitation or an inhibition of thermoreceptors (e.g., an increase in heat loss may be due to an activation of warm receptors or an inhibition of cold receptors during skin heating), no conclusion can be drawn as to which type of receptor is involved during temperature changes in this range.

1.1.7 Efferent Control of Thermoreceptor Activity

In isolated preparations of frog cold afferents an enhancement of dynamic sensitivity was observed after sympathetic stimulation and after perfusion with epinephrine or norepinephrine, the latter effect being reduced after application of α- and β-blocking agents (Spray 1974a). In primate cold receptors there was no effect of α- and β-blocking drugs when a fall in blood pressure was restored by dextran (Iggo and Paintal 1977). At a high body temperature the sympathetic activity is enhanced; this had no effect on the normal function of cold receptors, but the threshold of the paradoxical response was shifted to lower temperatures (Long 1977). Altogether, an efferent control of mammalian thermoreceptors seems to be questionable.

1.1.8 Long-Term Adaptation

An important question for temperature regulation ist whether there is acclimation of thermoreceptor activity to temperature. In frog cold receptors acclimated to 9.5 °C the dynamic sensitivity and the range of static activity increased whereas the static maximum decreased when compared with warm-acclimated (23 °C) animals (Spray 1975). In cats adapted to 5 °C there was no effect on cold receptor characteristics in cats after 1 or 2 years (Hensel and Banet 1978). After 4 years of adaptation cold receptor dynamic and static maxima shifted from 30° to 25 °C when compared with those of warm-acclimated animals (Hensel and Schäfer 1979). Since these experimental conditions do not correspond to natural conditions one might question whether there are seasonal variations in thermoreceptor sensitivity.
Since there is a circadian variation in body temperature it would be interesting to see whether there is also a circadian variation in thermoreceptor sensitivity, but no electrophysiological investigations have yet been done.

1.1.9 Temperature-Sensitive Mechanoreceptors

There are two types of slowly adapting mechanoreceptors which are sensitive to thermal stimulation. Type I responds to thermal stimuli only when mechanically activated. There is an excitatory overshoot during rapid cooling, but not with slow rates of temperature changes. The static curves show a maximum near 40 °C, which is true both for mammals (Duclaux and Kenshalo 1972) and birds (Necker 1973; Necker and Reiner 1980). Type II often has spontaneous activity in the range of normal skin temperatures and shows a phasic response to rapid cooling (Chambers et al. 1972); the static maximum was found to be near 30 °C in cats. This type of mechanoreceptor could contribute to thermal information from the skin. This seems to be unlikely, however, since these mechanoreceptors are innervated by thickly myelinated fibers. Moreover it has been shown by nerve block experiments in man that thermal sensation can be blocked differentially without blocking mechanical sensation due to conduction in large fibers (Fruhstorfer 1976). In addition, radiant thermal stimulation, which is a powerful stimulus for thermoreceptors, is ineffective in eliciting a response in mechanoreceptors (Duclaux 1977; Necker and Reiner 1980). In this way a contribution of mechanoreceptors to thermal information may be doubted.

1.2 Thermoafferent Pathways

1.2.1 Anatomical Outline

The axons of cutaneous thermoreceptors synapse in the dorsal horn of the spinal cord or in the spinal trigeminal nucleus (trigeminal, facial, glossopharyngeal, and vagal fibers; Nieuwenhuys et al. 1978). Neurons which respond specifically to thermal stimuli seem to be concentrated in the marginal layer (lamina I of Rexed 1952) of the dorsal horn or the caudal part of the spinal trigeminal nucleus, which shows a lamination similar to that of the dorsal horn. The axons of these neurons ascend in the spinothalamic tract (Trevino and Carstens 1975; Trevino 1976; Willis et al. 1979; Boivie 1979) or in the lateral trigeminothalamic tract (Nieuwenhuys et al. 1978). Whether the thermoafferent pathway is restricted to the anterolateral tract of the spinal cord may be doubted since thermal discrimination in monkeys (Eidelberg and Rick 1975) and cats (Norrsell 1979) is not abolished after anterolateral lesions. "It appears that there is a large safety factor for the conduction of the sensory afferent data signalling skin temperature" (Eidelberg and Rick 1975).

Spinothalamic tract axons terminate in the posteromedial nucleus and the posterolateral ventral nucleus of the thalamus (ventrobasal complex; Boivie 1979). Part of the axons of the anterolateral tract or collaterals terminate in the reticular formation (spinoreticular tract), the raphe nuclei, or the central gray of the midbrain. Spinoreticular fibers are the first link in a multisynaptic pathway that terminates in the intralaminar nuclei of the thalamus (Nieuwenhuys et al. 1978).

It is not yet clear whether thermal information from the skin is transmitted to the hypothalamus via the spinothalamic pathway or via the multisynaptic medial brain stem pathway.

1.2.2 Specific Responses of Central Neurons

There is generally a high degree of convergence in central neurons of the somatosensory pathway. In addition to large receptive fields which result from convergence of many receptors of the same type onto a single neuron, there is often a convergence of receptors of different sensory modalities; e.g., there are neurons which can be activated by mechanical and thermal and painful stimulation of the skin. The question arises whether only neurons which respond specifically to thermal stimulation are involved in the transmission of skin temperature or whether neurons which have a clear input from thermoreceptors, but in addition also from mechanoreceptors or nociceptors, may serve the same function. This problem has not been solved at the moment.

There is a continuous transition from neurons responding to light up to strong or noxious mechanical stimulation in addition to the response to thermal stimulation. Furthermore, the response to thermal stimulation may result from an activation of temperature-sensitive mechanoreceptors. The response to very strong mechanical or noxious thermal stimulation may result from an activation of thermoreceptors or from convergence from nociceptors. In this way, especially when including noxious stimulation, no clear distinction between specific and nonspecific neurons can be made. On the other hand, the responses of nonspecific neurons to thermal stimulation do not differ fundamentally from those of specific neurons. This means that the central processing of thermal information could also be studied in nonspecific neurons responding to small nonnoxious temperature changes.

1.2.3 Dorsal Horn Neurons

Dorsal horn neurons responding to innocuous thermal stimulation of the tail skin were first described in the marginal zone of the dorsal horn of cats (Christensen and Perl 1970). Most units were excited by cooling the skin and some by warming. Most units were additionally activated by intense mechanical manipulation of the skin, but not by moderate mechanical stimuli. Cold units often showed a response to noxious heating comparable to the "paradoxical" response of cold receptors. Similar qualitative results were found in the dorsal horn marginal zone of the cat and monkey (Iggo and Ramsey 1974, 1976; Kumazawa et al. 1975; Kumazawa and Perl 1977, 1978). Static curves were not assessed in these investigations.

Burton (1975) studied dorsal horn neurons in the cat and monkey which responded to cooling and warming steps and to constant temperatures applied to the hindlimb. These neurons were distributed throughout the dorsal horn. All responded to mechanical stimuli, but some only to noxious ones. Receptive fields were large, exceeding 20 mm^2. There was a linear increase in dynamic response to cooling steps (adapting temperature 35 °C) down to 15 °C in some neurons, but in others there was a steep increase within a narrow temperature rang (about 35 ° – 30 °C) with no further increase at lower temperatures. Often the same neurons showed an increase to warming steps starting at 35 °C, which was nearly linear up to 49 °C. Under static conditions no significant dependence of firing rate on skin temperature could be observed, i.e., the responses were mainly phasic.

In another quantitative investigation where only radiant heat was applied to the hindlimb of cats, again all temperature-responsive neurons were distributed throughout the dorsal horn and responded to mechanical stimuli, but some only to noxious stimuli (Price and Browe 1975). Warm unit thresholds ranged from 36 ° to 40 °C, and peak response usually occurred at 40 ° – 41 °C. A second group of dorsal horn neurons called "warming-noxious" units had thresholds between 38 ° – 42 °C, but the response increased up to about 45 °C and then plateaued.

In a quantitative study which excluded noxious stimuli dorsal horn neurons responding to thermal stimulation of the scrotum of the rat were investigated. Besides neurons which responded to both mechanical and thermal stimulation, neurons were found which responded exclusively to thermal stimulation (Hellon and Misra 1973a). They were located throughout the dorsal horn. Receptive fields were very large, including the contralateral side (Hellon and Mitchell 1975). Units responding to warming and units responding to cooling were found. Some neurons showed a dynamic response and a static response, some only a dynamic and others only a static response. Static curves of warm units had thresholds between about 27 ° and 40 °C and an increased firing rate up to 43 °C. Some warm units showed a steep increase in firing rate within a narrow temperature range ("switch-type" response) and then plateaued; the threshold of such neurons could be modified by separate thermal stimulation of the contralateral scrotal skin (Hellon and Mitchell 1975), indicating a high degree of processing at the spinal level in this case. The latter response type was found to dominate in the thalamus (Hellon and Misra 1973b). Static curves of cold units showed an increase of activity below 40 °C (Hellon and Misra 1973a) down to as low as about 15 °C, but some had maxima near 30 °C.

1.2.4 Trigeminal Nucleus Caudalis Neurons

The caudal portion of the spinal trigeminal nucleus caudalis which extends from the obex to the first cervical segments of the spinal cord shows a lamination similar to that of the spinal cord dorsal horn. Again, most neurons responding specifically to thermal stimulation of the skin of the face were found in the marginal layer. Receptive fields in the cat were generally between 10 and 100 mm^2, with only a few smaller or larger (Fruhstorfer and Hensel 1973; Dostrovsky and Hellon 1978), and all were ipsilateral.

Cold units insensitive to mechanical stimulation have been demonstrated in the cat (Mosso and Kruger 1973; Fruhstorfer and Hensel 1973) and monkey (Price et al. 1976). These neurons were sensitive to small changes in temperature.

In the cat and monkey, nucleus caudalis neurons responding specifically to thermal stimuli of the tongue were studied (Poulos and Molt 1976). Cold units were sensitive to small temperature changes and showed dynamic rate increases in response to increasing stimulus intensities. Mean static curves showed an increase in firing rate below about 41 °C and a maximum between about 30 ° and 25 °C, in this way being similar to lingual cold receptors (Poulos and Lende 1970a, b).

In a quantitative study of facial temperature representation in the cat many neurons were found in the marginal zone of the nucleus caudalis which responded to cooling of the face; a few responded to warming (Dostrovsky and Hellon 1978). Some neurons were weakly activated by mechanical stimulation. Noxious stimulation was avoided. There was no difference between neurons that had a weak mechanical input and those without such an input. The burst pattern of peripheral cold receptors was not found in the nucleus caudalis neurons. In cold units rapid cooling was followed by a phasic response and warming by a transient inhibition comparable to the response of cold receptors. All static curves showed an increase in firing rate below 40 °C (the upper experimental temperature limit in these experiments) and maxima between about 20 ° and 30 °C, which is again very similar to peripheral receptors (see Fig. 2). Warm units showed phasic responses and static curves with an increase of firing rate above 35 °C and maxima between about 40 ° and 45 °C or higher (temperature limit 45 °C).

Similar results were found when this investigation was extended to rabbits and rats (Dickenson et al. 1979).

1.2.5 Ventrobasal Complex of the Thalamus

Thalamic neurons specifically responding to thermal stimulation of the tongue were described by Landgren (1960), Emmers (1966), Poulos and Benjamin (1968), and Poulos and Molt (1976). In the quantitative investigation of Poulos and Benjamin (1968), both excitatory and inhibitory phasic responses were found in cold units (no warm units were found). Static curves showed an increase in firing rate with decreasing temperatures below 45 °C and reached a flat maximum near 30 °C. Altogether, the response characteristics were very similar to those of peripheral receptors, including bursts (which could also be an inherent feature of these neurons), so that little processing of information seems to occur.
Neurons responding specifically to thermal stimulation of the scrotal skin in rats were found predominantly in the ventrobasal complex (Hellon and Misra 1973b; Jahns 1975). The "switch-type" response, i.e., a steep increase in firing rate within a narrow temperature range (see Fig. 3), which was occasionally observed in the dorsal horn, was found nearly exclusively in the thalamus. The operating temperatures of all neurons were within the range 31 ° – 40 °C. Most neurons were excited by warming and some were inhibited by warming (or excited by cooling; there was no dynamic or phasic response). Since both types were found in the very same temperature range correspondings to that in which both cold receptors and warm receptors are sensitive, it is impossible to say whether the response was due to one or the other peripheral receptor type. When the number of activated warm units was examined, an increase with increasing skin temperature up to about 38 °C was found (Fig. 3). It has been suggested that in this way an increase in skin temperature could be signaled by an increase in the number of excited neurons (Hellon and Misra 1973b). Jahns (1975) analyzed the firing pattern of warm units and found that, at low skin temperature, thalamic neurons often changed to a bursting pattern.

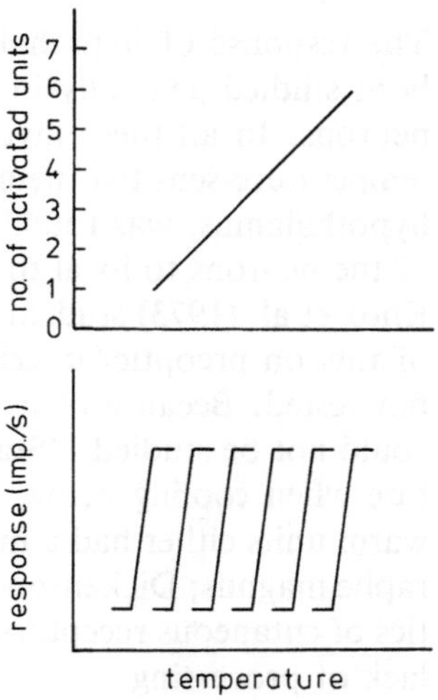

Fig. 3. Response of thalamic neurons to thermal stimulation of the scrotal skin (see Hellon and Misra 1973b). The *lower diagram* schematically shows the typical switch-type response of six different units with different thresholds. From this it follows that there is an increasing number of activated units with increasing temperature as shown in the *upper diagram*

Thalamic neurons responding to cooling, but in addition to mechanical stimulation of hand and foot skin, were found in the squirrel monkey (Burton et al. 1970). Both cold and warm units which were again activated by mechanical stimulation and which had receptive fields all over the body surface were found in the cat (Martin and Manning 1971).

1.2.6 Cortex

There are several investigations of cortical neurons in the somatosensory areas which respond to thermal stimulation of the skin (Landgren 1957; Kreisman and Zimmermann 1973; Hellon et al. 1973), but since there is no evidence for an involvement of these areas in temperature regulation, no detailed description shall be given here.

1.2.7 Lower Brain Stem

Neurons which were excited specifically by warming the scrotal skin in rats were found in the raphe nuclei (most in the nucleus raphes medianus) and in the central gray substance of the midbrain (Jahns 1976). During overall thermal stimulation of the skin of the trunk in rats, Dickenson (1977) found both cold units and warm units (most in the n.r. magnus of the medulla). No dynamic responses were observed, but the rate of temperature change was rather slow. The static activity of cold units increased below about 35 °C and peaked at about 29 °C in the mean curve. Warm units showed a clear increase of activity above about 35 °C and a peak at about 38 °C. Beyond the maximum there was always a clear decrease in firing rate. Inoue and Murakami (1976) found warm units and cold units in the medullar reticular formation which responded to skin stimulation, but no detailed description was given.

1.2.8 Hypothalamus

The response of hypothalamic neurons to thermal stimulation of the skin has been studied generally in addition to a study of the thermosensitivity of such neurons. In all these investigations an influence of skin temperature on some temperature-sensitive neurons, both in the preoptic area and in the posterior hypothalamus, was found which in some way or another modified the response of the neurons to local thermal stimulation (see Sect. 3.4).

Knox et al. (1973) studied the effect of radiant heating or cooling of the tail skin of rats on preoptic/anterior hypothalamic neurons. Mechanical sensitivity was not tested. Because of a slow rate of temperature change the phasic response could not be studied. "Static curves" of cold units showed an increase of firing rate when cooling below 40 °C and a maximum near 30 °C. "Static curves" of warm units either had a maximum near body temperature (cf. warm units in the raphe magnus; Dickenson 1977) or at or above 40 °C. In this way the characteristics of cutaneous receptors are also represented in the hypothalamus, indicating a lack of processing.

The response of hypothalamic (anterior and posterior area) neurons to thermal stimulation of the scrotal skin was studied in detail in addition to local thermosensitivity of the neurons (Nakayama et al. 1979). Neurons sensitive to thermal stimulation of the hypothalamus and neurons not sensitive showed a similar response characteristic of peripheral stimulation which, on the other hand, closely resembled that of thalamic neurons (Hellon and Misra 1973b): The neurons were insensitive to mechanical stimulation and showed the switch-type response (increase of firing rate within a narrow temperature range, usually less than 4 °C), some being excited by warming and some inhibited. The operating range of these neurons was between 30 ° and 41 °C and in this way very similar to that of thalamic units. Receptive fields were bilateral. Some neurons were excited by warming, but increased their firing rate when cooling from 30 ° to 20 °C, which clearly points to an input from cold receptors.

1.2.9 Central Processing: Thermoregulatory Aspects

In comparing the scrotal thermal afferent pathway and the trigeminal system, which have been studied in the most detail, one finds quite different modes of central processing. In the scrotal pathway there is considerable processing with a high degree of spatial integration and a conversion of the bell-shaped curves of warm receptors into a switch-type response with a narrow operating range. This processing seems to be completed even at the level of the spinal cord. Taking into account that the scrotal thermoafferent system probably serves a special function in reproduction the question arises whether it can be considered as being typical for skin temperature processing for thermoregulation. Heating the scrotal skin activates not only heat loss mechanisms (Waites 1962), but also vasomotor responses (Neya and Pierau 1976). "The scrotal pathway is evidently not a good model for describing the processing in thermal pathways from skin areas generally" (Dostrovsky and Hellon 1978).

In the trigeminal system there is a lack of central processing. The characteristics of the central neurons responding to thermal stimulation of the skin resemble those of the peripheral thermoreceptors.

Static curves of neurons responding to other skin areas were studied in less detail, but most had characteristics comparable to the peripheral receptors. Altogether, the characteristics of peripheral receptors seem to be represented at all levels of the thermoafferent pathway so that the cautious conclusion may be drawn that, except for the scrotal system, at the level of transition from the afferent to the efferent part of the thermoregulatory system the response characteristics of the neurons resemble those of the skin receptors. Whether warm units drive the heat loss mechanisms and cold units the heat gain mechanisms cannot be deduced from the electrophysiological experiments.

1.3 Thermoregulatory Responses to Thermal Stimulation of the Skin

1.3.1 General Considerations

The influence of skin temperature on the various thermoregulatory activities has been critically reviewed in detail by Bligh (1973). Only a short survey based on selected recent investigations will be given here.

Besides behavioral responses, which have been reviewed elsewhere (Corbit 1970; Cabanac 1979) and which will not be considered here, there are various autonomic responses which serve the adjustment of body temperature in homeotherms. In the thermoneutral zone, changes in the skin blood flow regulate heat loss or heat conservation by vasoconstriction or vasodilatation. A similar function, i.e., change in heat transfer from the body core to the environment, serves a change in insulation by pilomotor responses (Hensel et al. 1973). In a cold environment heat gain results from extra metabolic heat production by shivering. Nonshivering thermogenesis seems to be confined to neonates of mammals and seems to play a role in cold adaptation (Hensel et al. 1973). In hot ambient temperatures or during an increase in body temperature in exercise, heat loss is effected by evaporation, which is increased either by saliva secretion and increased respiratory rate (or saliva spreading, e.g., in the rat; Hainsworth and Stricker 1970) or by sweat secretion.

It has been shown that nearly all parts of the body are in some way or another sensitive to thermal stimulation with regard to thermoregulatory activity. From this it becomes evident that all other parts have to be kept constant or controlled when examining the influence of skin temperature on thermoregulation. Such a condition is difficult to achieve, and this problem has often been disregarded in studies on skin thermosensitivity so that the interpretation of some investigations must be considered with reservations.

The concept of the role of skin thermoreceptors on temperature regulation was influenced by the investigations of Benzinger in humans (1970). Benzinger concluded that only *cold receptors* of the skin play a role in temperature regulation while warm reception is confined to the body core. Bligh (1973) argued against this strict separation and recent investigations do not confirm it (see below).

During overall thermal stimulation of the skin, the temperature of insulated skin areas usually varies in the range 30°–40°C. In this very range both cold receptors and warm receptors are sensitive to thermal stimulation (see Sect. 1.1 and Fig. 2). From this it seems to be hopeless to try to estimate whether an effect is due to a stimulation (or inhibition) of cold receptors or warm receptors.

Another problem concerns the different thermal sensitivity of different skin areas which might result from a varying density of thermoreceptors or different weighting during central processing. Usually a mean skin temperature is calculated by weighting different areas according to the assumed importance. There are only a few investigations of the differential thermal sensitivity of different skin areas, but such investigations are clearly very important and will be considered in a separate chapter (Sect. 1.3.3).

1.3.2 Overall Thermal Stimulation

Heat Production, Shivering. There is ample evidence that skin temperature influences heat production (for review see Bligh 1973; Hensel 1973a). At the normal core temperature, cooling the skin results in an increase in heat production. This has recently been confirmed and, in addition, it has been shown that skin warming also has an influence. The latter influence could be shown when an increased heat production was established by central cooling. Skin warming then resulted in a reduction of heat production. Such a condition was achieved by immersion in a water bath in man (Hayward et al. 1977) and in rats (Morhardt et al. 1975), by cooling the spinal cord in pigeons (Rautenberg 1971) and in rats (Banet and Hensel 1976), by cooling the hypothalamus in rabbits (Boulant and Gonzales 1977), and by cooling the hypothalamus and the spinal cord in pigs (Carlisle and Ingram 1973) and goats (Jessen 1977). In the water bath skin temperature was that of the water and ranged from 28° to 40°C in the rat (Morhardt et al. 1975) and 12° to 40°C in man (Hayward et al. 1977). In the pig, trunk skin was changed by a coat and had an effect in the temperature range 25°–40°C (Carlisle and Ingram 1973). In the rabbit effective skin temperature ranged from 33° to 40°C at different ambient temperatures (Boulant and Gonzales 1977). The relationship between skin temperature and heat production was linear up to 40°C in these experiments.

Respiratory Evaporative Heat Loss, Panting. It seems to be difficult to elicit panting or an increase in breathing rate by an increase in skin temperature at the normal core temperature. This holds true for man, mammals, and birds. But after a slight increase in internal temperature, which may be confined to the hypothalamus or to the spinal cord, heating the skin results in an increase in respiratory rate, and at higher core temperatures cooling the skin is followed by a decrease. Recently in rabbits, back skin warming (Stitt 1976) or thermal stimulation of the trunk skin (Boulant and Gonzales 1977) in the skin temperature range 33°–40°C was shown to influence respiratory evaporative heat loss in combination with hypothalamic thermal stimulation.
In pigs (Ingram and Legge 1972) and in goats (Jessen 1977), heat loss induced by hypothalamic and spinal heating was modified by skin temperature or ambient temperature changes. In pigeons similar results were found during spinal thermal stimulation (Rautenberg et al. 1978). In chickens radiant thermal heating of distinct skin areas was not followed by an increase in respiratory frequency until rectal or hypothalamic temperatures rose (Richards 1970, 1971).

Sweating. The role of skin temperature in sweating has been predominantly studied in man. Although there is a local thermal effect on the sweat gland (Hensel 1973a), a clear reflex response can be shown by eliciting sweat secretion in an area remote from the stimulated area (Bullard et al. 1970; McCook et al. 1970). Similar to panting, sweating can be influenced by skin stimulation usually only after an increase in the internal temperature. Under these conditions heating the skin with heat lamps was followed by an increase in the sweating rate over a skin temperature range of 34°–36°C at constant core temperature (Nadel et al. 1971). This result was confirmed by McCaffrey et al. (1979) for a temperature range of 33° – 40 °C where a linear relationship was found between skin temperature and sweating rate at a constant core temperature. This shows that both cold and warm skin do influence sweat secretion. Wyss et al. (1974) were not able to demonstrate a significant influence of skin temperature, but in their experiments both skin and core temperatures were changed simultaneously, starting at low core temperatures; thus, a clear statement of the effect of skin temperature change cannot be given.

Skin Blood Flow. Changes in skin blood flow usually result in a change in skin temperature, but this is only an indirect indication which by no means gives a quantitative measure of blood flow. Here only investigations are included in which skin blood flow was measured. As with sweating there is a direct effect of local temperature on skin blood flow, but a reflex control which involves peripheral thermoreceptors is also observed (for review see Keatinge 1970; Bligh 1973; Hensel 1973a). Skin blood flow also depends on internal temperature, but the effects of skin temperature stimulation can be seen at normal core temperature. In the pig, temperature changes of the trunk skin resulted in a change in tail blood flow which was linear in a skin temperature range of about 30° – 40 °C under conditions in which spinal and hypothalamic temperatures were controlled (Ingram and Legge 1972). Similar results were found in monkeys (Proppe et al. 1976; Proppe 1978; Lynch et al. 1980). In investigations in man skin and core temperatures were changed simultaneously (Wyss et al. 1974; Johnson and Park 1979), and it was concluded that skin temperature influences skin blood flow. This method, however, does not allow a clear correlation between skin temperature and blood flow.

There are several investigations on foot skin blood flow in birds (e.g., Johansen and Millard 1973; Bernstein 1974), and the feet in birds are clearly very important for heat loss (Steen and Steen 1965). Since thermal stimulation of skin and body core were not controlled separately in these investigations, no clear correlation between skin temperature and blood flow can be established.

1.3.3 Thermal Stimulation of Selected Skin Areas

Thermal stimulation of selected small skin areas usually does not influence other parts of the body and thus provides a more unambiguous indication of skin sensitivity. In addition, the sensitivity of different skin areas can be compared. On the other hand, the thermoregulatory responses are often small, and this might be the reason why only a few investigations with selected skin stimulation have been reported.

The most impressive results were found with scrotal skin stimulation in rams (Waites 1962; Hales and Hutchison 1971) and pigs (Ingram and Legge 1972). Heating the scrotal skin to 40° – 42 °C, which is in the range in which only warm receptors are activated (Hellon et al. 1975), was followed by an immediate increase in respiratory frequency which was maintained despite a fall in core temperature by as much as 2 °C. In the long run there was a stabilization of core temperature on a lower level (Hales and Hutchison 1971). This response could be inhibited by cooling other skin areas (e.g., trunk). Heating a similar area of the body trunk did not elicit a significant response, which shows that the trunk skin is much less sensitive than the scrotal skin (but the larger area of the trunk skin may compensate for the low sensitivity).

Scrotal heating influenced not only respiratory evaporative heat loss, but also skin blood flow. There was an increase in tail blood flow during scrotal heating even at low trunk skin temperature at which no increase in respiration rate was observed. There was again a fall in core temperature (Hales and Hutchison 1971; Ingram and Legge 1972). At low skin temperatures there was an increased heat production which was reduced by scrotal skin heating (Hales and Hutchison 1971; Ingram and Legge 1972).

Neya and Pierau (1976) showed in the rat that there was an increase in blood pressure which had a threshold similar to that of the switch-type response of dorsal horn or thalamic thermoresponsive neurons (see Sect. 1.2). This points to a special role of the scrotal thermoafferent system in that heating might result in an alarm reaction and not only in thermoregulatory responses.

Heating and cooling the ears and the back skin and their influence on metabolic heat production have been studied in the rabbit (Kluger et al. 1972). The temperature of ear skin was changed in the range 10 ° – 30 °C, which is only in the sensitivity range of cold receptors. A change in ear skin temperature was followed by a change in heat production. Heating the back skin from 30 ° to 37 °C resulted in a reduction of heat production. No significant change of core temperature was observed during these stimulations. The sensitivity was very similar in both areas (similar response after correction of stimulus and area). When comparing this selective stimulation with total animal stimulation, the heat production response was greater, as one would expect. This is probably also true for scrotal stimulation where heat loss is high despite a considerable fall in other body temperatures. This points to a nonsummative integration of temperature signals from different skin areas.

The effect of thermal stimulation of selected skin areas on ongoing shivering due to a low body temperature was tested in lightly anesthetized pigeons. Heating the naked parts of the feet and the beak was rather ineffective in influencing shivering when the skin temperature was increased to about 45 °C (Necker 1977). This effect is in contrast to electrophysiological results which showed that there were numerous cold receptors in the beak (Necker 1972). In contrast to this, feathered skin areas were very sensitive to heating from about 35 ° to 43 °C in the absence of any core temperature change. Cooling the skin below 35 °C resulted in a increased shivering. When comparing approximately equal areas, back skin was most sensitive, wing skin less sensitive, and the skin above the breast muscle least sensitive. In addition there was an interaction between different skin areas, as found with scrotal heating in the ram and pig.

Differential thermal sensitivity was studied with regard to the effector response of sweating in man (Nadel et al. 1973; Crawshaw et al. 1975). The sensitivity of the face was about three times that of the thighs, the chest, and the abdomen, and that of the lower legs half that of the latter ones when comparing the sweating rate elicited by radiant heating of discrete areas (skin temperature about 35 ° – 40 °C) corrected for the area and the stimulus intensity (Nadel et al. 1973). The sweating rate in a warm environment was reduced by cooling selected areas (Crawshaw et al. 1975) from about 36 °C to 16 ° – 22 °C. A similar differential thermal sensitivity was found as in the previous investigation (Nadel et al. 1973). In addition, psychophysical measurements were made which showed that the stimulus magnitude estimate correlates well with the thermal sensitivity with regard to the thermoregulatory response. This indicates that sensory discrimination and thermoregulation obviously receive a similar input which might depend on the distribution or area density of thermoreceptors.

1.3.4 Conclusion

There is now considerable evidence that both cooling *and* warming the skin influence thermoregulatory activity. In this respect, Benzinger's concept that only skin cold receptors are involved in thermoregulation (Benzinger 1970) has not been supported by recent investigations when one takes into account that in the skin temperature range of 35° – 40°C both warm receptors and cold receptors are sensitive to thermal stimulation. A clear demonstration of the activation of skin warm receptors only and their effect on thermoregulation is given by the experiments using scrotal skin heating in the range of maximum warm receptor activity.

2 Deep-Body Thermosensitivity (Excluding CNS)

2.1 Deep-Body Thermoreceptors

In cutaneous thermoreceptors the problem of the specific function is confined to an additional mechanical sensitivity. In enteroreceptors, besides mechanical sensitivity, a variety of chemical substances can be adequate stimuli. Consequently, many different substances should have been tested to prove the specific function, but because this has not been done as yet, the specific function of temperature-sensitive enteroreceptors has not been shown beyond any doubt. In addition, deep-body receptors whose specific function is known have been shown to be temperature sensitive: chemoreceptors of carotid bodies are highly warm sensitive, having a mean Q_{10} of 75 (Gallego et al. 1979); muscle spindle activity is warm sensitive in most cases, especially near body temperature (Mense 1978). Thin fiber (group III and group IV) afferents of muscle nerve have been demonstrated to be warm sensitive or cold sensitive without responding to mechanical stimuli. However, some of these fibers are excited by bradykinin, a pain-producing substance (Hertel et al. 1976).

Warm-sensitive receptors which were not excited by mechanical stimuli were found in the abdominal wall of rabbits (Riedel 1976). As with cutaneous warm receptors two types could be distinguished: one type had its maximum near 40°C and the other type near 46°C. Again as with cutaneous receptors, the dynamic sensitivity was maximal near the static maximum. Cold-sensitive afferents in the splanchnic nerve of cats were excited by low temperatures (thermode temperature 12°C), but were insensitive to mechanical stimulation (Gupta et al. 1979). Vagal cold and warm receptors which have been shown to be involved in gastroduodenal motility could be demonstrated in the cat (El Quazzani and Mei 1979). Cold receptors showed no spontaneous activity at 37 °C and were excited again at very low temperatures (10 ° – 12 °C). Warm receptors had maxima near 45 °C.

2.2 Thermoregulatory Responses to Deep-Body Thermal Stimulation

In sheep intraabdominal heating was shown to influence thermoregulatory activity, e.g., an increase in respiratory frequency (Rawson and Quick 1970, 1972). Since other body temperatures were controlled it could be ruled out that the response was not due to stimulation of other areas. Ruminal cooling to 16 °C was followed by an immediate fall in respiratory frequency and subsequent shivering in a thermoneutral environment (Rawson and Quick 1972) so that not only heating, but also cooling turned out to be effective. It was concluded that the thermoreceptors should lie in the wall of the rumen and intestine, but not in the posterior vena cava. In experiments in rabbits the dorsal wall of the abdominal cavity was stimulated selectively and other body temperatures controlled (Riedel et al. 1973). Heating was followed by an increase in respiratory rate. Cooling was less effective and led only to a slight decrease in breathing rate when cooling to 36 °C; no further decrease was observed when cooling to as low as 10 °C. From this it was concluded that only warm receptors are involved. Accordingly, integrated afferent activity of the splanchnic nerve showed a change in activity only above about 36 °C (Riedel et al. 1973); in addition, the response to heating can now been explained by the demonstration of warm receptors (Riedel 1976; see Sect. 2.1).

From experiments in rabbits where thermal stimulations (heating and cooling) of the circulatory system were performed, it was concluded that there are no temperature sensors in the right heart, the lung, the posterior vena cava, the posterior vein, the liver, or in the hepatic vein (Cranston et al. 1977, 1978). This is in partial agreement with the experiments of Rawson and Quick (1972) in sheep.

In goats, where an independent thermal control of hypothalamus, spinal cord, and "residual inner body" was achieved by intravascular heat exchangers, both warming and cooling were effective, and it was shown that cold sensitivity was comparable to that of the hypothalamus and spinal cord (Mercer and Jessen 1978). No localization of receptor sites was tried. In a preliminary report it was shown that extra central nervous deep-body cold sensitivity in ducks is higher than that of brain or spinal cord (Simon and Simon-Oppermann 1979). These experiments clearly show that there is an important deep-body thermosensitivity outside the central nervous system.

3 Thermosensitivity of the Central Nervous System

3.1 Spinal Cord

3.1.1 Thermoregulatory Responses to Thermal Stimulation

There is now ample evidence that cooling and warming the spinal cord via extradural thermodes in the vertebral canal are followed by appropriate thermoregulatory responses: cooling results in shivering, reduced skin blood flow, and pilomotoric response, while heating results in increased respiratory frequency, increased skin blood flow, and increased sweating rate. In addition, heat loss mechanisms are inhibited by spinal cooling and heat gain mechanisms by spinal warming. These effects have been shown both in mammals and in birds and have been reviewed in detail by Simon (1974) and in further reviews which include this topic (Hellon 1972b; Bligh 1973; Hensel 1973a; Cabanac 1975).

Thermal stimulation of the spinal cord has been done recently, mainly in the context of an evaluation of the interaction of various temperature-sensitive sites which will be dealt with separately (see Sect. 3.4). Since these investigations do not give new insights into spinal thermosensitivity they will not be treated here. This section will concentrate on recent investigations which deal mainly with the importance of spinal thermosensitivity.

Responses in Spinalized Animals. Spinal transections at a high cervical level do not totally inhibit thermoregulatory responses to thermal stimulation of the spinal cord. This has been interpreted to show that thermoregulation can function even at the spinal level (Simon 1974). Spinal cooling in acute or chronically spinalized dogs (Simon et al. 1966), rabbits (Kosaka and Simon 1968), cats (Herdman 1978), and pigeons (Görke and Pierau 1979) was followed by shivering. The threshold of response was usually at a rather low level of spinal temperature in acute preparations, but could be similar to that of intact animals in chronic experiments (Herdman 1978; Görke and Pierau 1979). Shivering intensity was reduced, but one has to keep in mind that the spinal transection is a dramatic trauma to the CNS. The effect might be due to a direct activation of spinal motoneurons (Klussmann and Pierau 1972; see below).

Effects of Peripheral Denervation (Deafferentation). In dogs (Meurer et al. 1967; Jessen and Simon-Oppermann 1976) and in pigeons (Necker and Rautenberg 1975) it has been shown that, with thermal stimulation of the spinal cord peripherally deafferented by cutting the dorsal roots, both heat gain (shivering) and heat loss (vasodilatation, panting) could be activated. In pigeons (Necker and Rautenberg 1975) it could be shown that after peripheral denervation, body temperature is regulated at a slightly lower level in a cold environment. Clamping the spinal cord to normal values resulted in a poikilothermic state in a cold environment (there is no cold sensitivity of the hypothalamus in pigeons). This clearly demonstrates the ability of the spinal cord to function as a temperature-sensitive area in the control of body temperature.

Natural Fluctuations in Spinal Temperature. The consequence of temperature regulation in homeotherms is a relative constancy of the internal temperature. Since the spinal cord is located in the body core only small fluctuations of spinal temperature can be expected, and actually do occur. This seems to be in contrast to rather large deviations during thermal stimulation although relatively small changes may elicit a response when effector activity is primed by stimulation of another area (see Fig. 5 and Sect. 3.4). The question arises whether natural fluctuations of spinal temperature can be correlated with changes in thermoregulatory effector activity.

This problem has been studied so far only in pigeons (Graf and Necker 1979). The variation of deep-body temperature depends on the size of animals and increases with decreasing size. Variations of up to 1 °C often occur in pigeons. During continuous recordings at an ambient temperature of 20 °C a mean standard deviation of ±0.5 °C was found in the active period (light condition). When correlating the electromyogram, which is an indication of shivering (and in this way of heat production), no significant correlation with the spinal temperature was found at constant ambient temperatures of 10° and 20 °C, where shivering can be observed. (As argued above, an active effector system might be necessary to show an effect of spinal temperature change.) This seems to be discouraging (and corresponds to results in the hypothalamus; see Sect. 3.3), but probably results from the interaction of many temperature-sensitive sites which are not under control and which may change the importance of the spinal cord in the closed loop system.

While there was no clear correlation at constant ambient temperature, there was a correlation during slow ambient cooling: Shivering was correlated positively with the spinal temperature, i.e., due to the increased heat production in a cold environment there was an increase in deep-body temperature (Graf and Necker 1979). Even with rapid changes in ambient temperature no fall in spinal temperature was ever observed. This shows that sensory function of the spinal cord is difficult to demonstrate in experiments with natural variations of spinal temperature.

Day-Night Variation of Spinal Thermosensitivity. According to the well-known day-night variations of body temperature (Aschoff 1970), there must be a diurnal variation in spinal temperature. Such diurnal variations have been studied systematically only in pigeons (Graf and Necker 1979; Graf 1980a). As to the thermal sensitivity of the spinal cord there is a shift in threshold in that a stronger cooling stimulus has to be applied during the night than during the day to get a similar response in heat production (Graf 1980b). As to heat loss (panting), a *smaller* warming step is sufficient during the night to elicit a panting response comparable to that observed during the day. In this way the sensitivity to cooling seems to be reduced during the night and that to warming seems to be enhanced. This shows that there is not a general reduction in sensitivity during the night, as one might expect. No systematic investigation (selected thermal stimulation) of a shift in skin thermosensitivity in the night has been done so far.

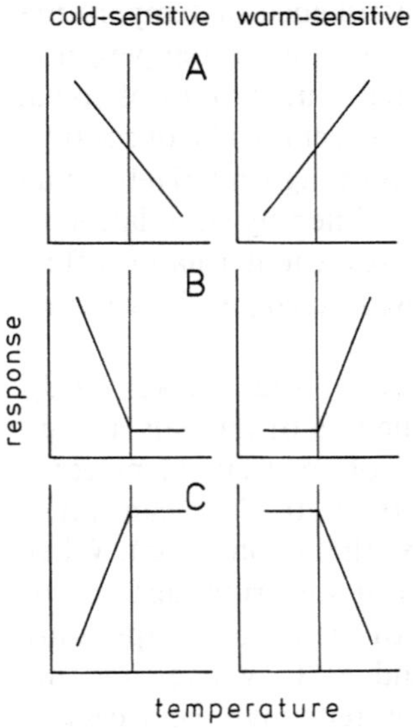

Fig. 4A – C. Schematic representation of types of responses of CNS neurons to thermal stimulation (not included: bell-shaped curves, see Fig. 2). **A** Linear response. **B** Threshold excitation response. **C** Threshold inhibition response. *Ordinate,* response (firing rate); *abscissa,* CNS temperature. *Vertical bar* shows normal value

3.1.2 Electrophysiological Evidence of Spinal Thermosensitivity

Only a few electrophysiological investigations, none recent, have examined the thermal sensitivity of the spinal cord in mammals (for detailed review see Simon 1974). There are two lines of electrophysiological approach. The first concentrated on the recording of afferent activity in ascending pathways. The second consisted of an investigation of the thermal sensitivity of motoneurons which are part of the efferent systems of thermoregulation.

In the CNS one has to distinguish between neurons which respond to thermal stimulation of a remote area, e.g., the skin, and which are in this way interneurons (they are named “cold units” or “warm units” in Sect. 1) and neurons which respond to local thermal stimulation. The latter will be named “temperature-sensitive” (cold sensitive, warm sensitive) in the following sections.

In hypothalamic temperature-sensitive neurons a distinction between sensory neurons (“thermodetectors”) and interneurons which receive an input from sensory neurons was proposed (Eisenman 1972). The same problem arises with spinal and lower brain stem neurons, but will be discussed in detail in Sect. 3.3.

Figure 4 shows different types of responses of temperature-sensitive central neurons which were found in all temperature-sensitive areas of the CNS.

In recordings from fibers in the anterolateral tract of the spinal cord (see Sect. 1.2), neurons were found in guinea pigs which responded to heating the spinal cord caudal to the recording site (Wünnenberg and Brück 1970). In cats, two groups of neurons were found which had characteristics similar to peripheral thermoreceptors: one group was excited by heating and one group was excited by cooling the spinal cord (Simon and Iriki 1971). As with peripheral receptors (see Fig. 2), there was a separation of the temperature range of maximal static sensitivity. The linear range of static curves of warm-sensitive neurons extended in the mean from 36° – 41 °C and that of cold-sensitive neurons from about 40° to 32 °C. Results similar to those in cats were found in the pigeon (Necker 1975a):

both warm-sensitive and cold-sensitive neurons were found, but there was an overlapping of the range of sensitivity (see Fig. 4A) which extended from about 37 ° to 44 °C (normal spinal temperature, about 41 °C). Neurons with a distinct threshold (see Fig. 4) were seldom observed. In all spinal investigations there was a preponderance of warm-sensitive units. Phasic responses were less pronounced than in peripheral receptors.

These experiments confirm what one would have expected: an activation of spinal neurons during thermal stimulation of the spinal cord. The problem arises whether there are specific thermosensitive neurons which detect spinal temperature or whether there is a nonspecific activation of spinal neurons which leads to a casual but adequate activation of neurons involved in thermoregulation. Since this is a general problem of central nervous thermosensitivity this question will be dealt with in a separate section (see 3.6).

The results that there is shivering in spinalized animals during spinal cooling (see above) stimulated the investigation of the thermosensitivity of motoneurons (Klussmann and Pierau 1972). It has long been known that spinal motoreflexes are enhanced during cooling of the spinal cord (Brooks et al. 1955). This has recently been confirmed for pigeons (Görke et al. 1975; Görke 1980). In intracellular recordings from motoneurons, enhanced postsynaptic potentials (PSP) were found during cooling (Klussmann and Pierau 1972; Pierau et al. 1976). This increase in PSP was probably due to an increase in membrane resistance measured during cooling. Whether there is, in addition to this direct effect, an input from spinal temperature-sensitive neurons or even peripheral thermoreceptors as proposed by Klussmann and Pierau (1972), which in this way may constitute a complete thermoregulatory system at the spinal level, has still to be elucidated.

These electrophysiological investigations of spinal thermosensitivity show that both afferent and efferent systems are influenced. This has to be kept in mind when discussing the thermoregulatory responses to spinal (and also higher central structures) stimulation.

3.2 Lower Brain Stem (Medulla Oblongata and Midbrain)

3.2.1 Thermoregulatory Responses to Thermal Stimulation

In contrast to studies involving the spinal cord (and hypothalamus), only a few investigations have examined the thermosensitivity of the lower brain stem. In the rabbit, cooling the midbrain was followed by an increase of shivering and oxygen consumption (Hardy 1969) and heating inhibited shivering (Cabanac and Hardy 1969), but this effect was small compared with hypothalamic stimulation. In the cat midbrain cooling was followed by an increase in hypothalamic temperature and heating in a decrease (Cronin and Baker 1977b); these effects were probably due to a change in the peripheral vasomotor tone since comparable temperature changes were observed in the nasal mucosa. In ducks, cooling the caudal midbrain and rostral medulla by about 10 °C was followed by vasoconstriction and a slight increase in heat production and body temperature (Simon-Oppermann and Martin 1979). In chickens, heating the midbrain induced panting (Richards and Avery 1978).

In monkeys (Chai and Lin 1972), heating the medulla oblongata by a thermode in the fourth ventricle to 42 ° – 43 °C resulted in vasodilatation and an increase in respiratory frequency; this response was comparable to the response to spinal heating. Cooling the medulla to 32 ° – 33 °C produced vasoconstriction and a decrease in respiration rate. Interestingly, cooling the medulla elicited shivering of the jaws only, while spinal cooling was followed by shivering of the limbs. This points to a local effect on the efferent system (motor nuclei of the trigeminal nerve in the medulla) as demonstrated for the spinal cord (see above). In addition there were effects on the circulatory system and on the state of wakefulness. Similar results were found in the rabbit (Chai and Lin 1973), and in addition it was shown that decerebration does not alter the responses. This supports the assumption that there could be an effect of thermal stimulation, mainly on the efferent side of the thermoregulatory system, or a complete regulatory system at this level of the central nervous system (see also Sect. 3.1.2!). The thermal stimulation of the lower brain stem is followed by changes in deep-body temperature, and this effect could also be demonstrated in rats (Lipton 1973).

3.2.2 Electrophysiological Evidence of Lower Brain Stem Thermosensitivity

There are few earlier investigations of the thermosensitivity of neurons in the lower brain stem (Hensel 1973a) and only few recent ones. In the reticular formation of the medulla of rabbits neurons were found which responded to local thermal stimulation (Inoue and Murakami 1976). Both warm-sensitive and cold-sensitive neurons were found, but the cold-sensitive ones were more numerous, which is in contrast to all other areas of the CNS where warm-sensitive ones dominate. Part of the neurons showed a linear dependence of firing rate on temperature over a rather wide temperature range of about 35 ° – 40 °C, but part had distinct thresholds distributed between 30 ° – 40 °C and were sensitive only within a small temperature range (as shown similarly in Fig. 3). The ranges of sensitivity of both warm-sensitive and cold-sensitive neurons overlapped (Fig. 4). Similar results (again a preponderance of cold-sensitive neurons) were found in the rabbit midbrain reticular formation (Hori and Harada 1976a).

In the rabbit, medulla respiratory neurons were found and their thermosensitivity studied (Inoue and Murakami 1976). These neurons (expiratory or inspiratory) typically discharge in groups or bursts, and the intervals determine the respiratory frequency. Cooling the medulla resulted in an increase of the number of spikes per burst, but the burst frequency decreased. This might explain the decrease in respiratory frequency which depends on the burst frequency of respiratory neurons during medullar cooling (see above). In this way, again, a direct influence of thermal stimulation on the efferent side of thermoregulation may be discussed from the electrophysiological results (see 3.1.2).

The search for thermosensitive neurons in the midbrain concentrated on the raphe nuclei since these neuclei have been supposed to be part of the thermoafferent pathway (see Sect. 1.2). A high concentration of thermosensitive neurons was found in the caudal part of the medial midbrain of cats (Cronin and Baker 1976, 1977a). Most were warm sensitive and had linear static curves over a wide temperature range of about 32° – 42 °C (Fig. 4A), but some had bell-shaped curves (most of the cold-sensitive neurons). Similar results were found in the rat raphe nuclei (Hori and Harada 1976b).

The thermal sensitivity of warm-sensitive neurons in the lower brain stem seems to be significantly lower than that of spinal cord or hypothalamus (which is about equal): calculations from the data of Cronin and Baker (1977a) yield a mean sensitivity of about 1.5 impulses/s °C as compared with about 5 impulses/s °C in spinal cord or hypothalamus (see Simon and Iriki 1971). This might explain the lower thermal sensitivity of the lower brain stem (see above) with regard to thermoregulatory responses.

3.3 Rostral Brain Stem (Hypothalamus)

3.3.1 Thermoregulatory Responses to Thermal Stimulation

The hypothalamus has long been recognized as being involved in thermoregulation, and there are numerous reports which show that especially the preoptic area and the anterior hypothalamus (POAH) are sensitive to thermal stimulation (Bligh 1973; Hensel 1973a; Cabanac 1975). Cooling this area results in an activation of heat conservation or heat production mechanisms and heating in an increased heat loss. Though there are differences in the sensitivity with regard to different thermoregulatory responses and in different species, it can be concluded that the thermosensitivity of the POAH has been shown beyond any doubt in mammals.

In birds such an outstanding thermosensitivity seems not to exist, as has been shown by several recent investigations. In sparrows the thermal stimulation of the preoptic area resulted in a change in oxygen consumption and respiration rate (Mills and Heath 1972). The effects on heat production were relatively small and might be due to a change in activity of the birds (wing flapping), which increased with cooling. In pigeons (Rautenberg et al. 1972), Adélie penguins (Simon et al. 1976), chickens (Scott and van Tienhoven 1974), California quail (Snapp et al. 1977), and in ducks (Simon-Oppermann et al. 1978), no increase in heat production was observed during hypothalamic cooling. There was a significant effect on skin blood flow (vasodilatation and vasoconstriction) and a less pronounced effect on respiration rate, indicating a slight warm-sensitivity.

Besides appropriate responses which were homeostatic, responses were observed in birds during hypothalamic stimulation which were nonhomeostatic. In a hot environment cooling the hypothalamus could result in an increase in respiration rate in pigeons (Schmidt 1976) and ducks (Simon-Oppermann et al. 1978). On the other hand, in a cold environment shivering and heat production were reduced by hypothalamic cooling in penguins (Simon et al. 1976) and in ducks (Simon-Oppermann et al. 1978).

Only a few investigations have used selective thermal stimulation of the posterior hypothalamus. This hypothalamic area is thought to incorporate the origin of the efferent side of the thermoregulatory system (Hardy 1973; Cabanac 1975). In a recent investigation in goats (Puschmann and Jessen 1978) it was shown that selective cooling of the posterior hypothalamus was followed by a reduction in an increased heat production due to a cold environment or anterior hypothalamic cooling. This means that both the anterior and posterior hypothalamus are thermosensitive, but in an opposite direction. If the sensitivity of the posterior hypothalamic area were equal to that of the POAH, a total compensation of the effects should be expected with naturally occurring brain temperature changes. (This could explain the lack of correlation between natural variation of hypothalamic temperature and thermoregulatory activity; see below.)

These results show that both homeostatic and nonhomeostatic responses can occur during hypothalamic thermal stimulation. In addition, the failure to demonstrate a correlation between natural variations in hypothalamic temperature and thermoregulatory activity (Abrams and Hammel 1964, 1965; Rawson et al. 1965) and the fact that deep-body temperature actually does change during CNS stimulation [increase of body temperature (Tb) during cooling and decrease during warming] show that a true sensory function of the CNS has to be discussed despite a clear thermosensitivity of these areas.

3.3.2 Electrophysiological Evidence of Hypothalamic Thermosensitivity

Since the hypothalamus is assumed to have a dual function in thermoregulation, i.e., a sensory function and an integrative function, different types of single-unit responses to local thermal stimulation should be expected. Figure 4 summarizes schematically the main types of responses to be expected. The sensory neurons ("thermodetectors," Eisenman 1972) should respond linearly over a relatively wide temperature range (Fig. 4A). As with peripheral receptors, there might be a separation of the sensitivity range of cold-sensitive and warm-sensitive sensory neurons but such a clear separation was not found in the hypothalamus or generally in the CNS (except for the spinal cord of cats; Simon and Iriki 1971). The integrative function may be represented by neurons (interneurons) which exhibit a threshold behavior, since thermoregulatory responses are characterized by a threshold. There might be neurons which are cold sensitive, having a threshold near normal Tb, or neurons which are warm sensitive, again with a threshold near Tb (Fig. 4B); a reverse function (inhibition at the cold or warm side, Fig. 4C) is also conceivable when taking into account that there are multiple inhibitory systems in the CNS. All these types of responses shown in Fig. 4 have been demonstrated (Guieu and Hardy 1970; Hellon 1967). Generally, warm-sensitive neurons were more numerous than cold-sensitive ones. Threshold neurons seem to be as numerous as linear neurons.

There are many fewer thermosensitive neurons in the posterior hypothalamus than in the POAH. This may explain the minor thermosensitivity of this area as stated previously. Nevertheless, both linear and threshold-type neurons were found, the latter obviously being more numerous. Both warm-sensitive and cold-sensitive neurons occurred (Edinger and Eisenman 1970; Wünnenberg and Hardy 1972).

Recent investigations in rabbits (Boulant 1974), ground squirrels and rats (Boulant and Bignall 1973a), golden hamsters and guinea pigs (Wünnenberg et al. 1976), goats (Mercer et al. 1978), and ducks (Simon et al. 1977) agree with earlier results (Bligh 1973; Hensel 1973a).

Interestingly, the thermosensitivity of preoptic neurons in ducks was very similar to that in mammals despite only a slight thermosensitivity with regard to thermoregulatory responses (see above). This shows that thermosensitive neurons are not a sufficient criterion for assuming a thermoregulatory function of a CNS area under study.

A comparison of thermosensitive neurons in a hibernator and a nonhibernator was carried out in two investigations. No difference could be observed between ground squirrels (hibernator) and rats (Boulant ad Bignall 1973a), but thermal stimulation was not extended below 30 °C and no quantitative estimation of thermal sensitivity was made. When comparing the thermal sensitivity of preoptic neurons in guinea pigs and golden hamsters (hibernator), linear-type warm-sensitive neurons (Fig. 4A) in the guinea pig became inactive or insensitive below 30 °C whereas in the hibernator, activity continued down to about 10 °C, although sensitivity decreased below 30 °C (Wünnenberg et al. 1976). In addition, the firing rate was much higher in hamster neurons. This difference may reflect the adaptation of a species to heterothermy. It has to be mentioned that the experiments in hibernators have been done in the nonhibernating state.

3.4 Interaction Between Different Thermosensitive Areas

3.4.1 Thermoregulatory Responses to Combined Thermal Stimulation of Two Different Areas

It has long been recognized that there is an interaction between different thermosensitive areas, e.g., skin, body core, spinal cord, and hypothalamus (Bligh 1973; Cabanac 1975). From this it follows that one has to know the thermal state of all thermosensitive areas when looking at the response to thermal stimulation of one selected area. (This is true also for selected thermal stimulation within one region!)

When the interaction between different thermosensitive areas was quantitatively studied, different types of response ("laws") were described. Figure 5A depicts a parallel shift of the response curves at different extrahypothalamic temperatures (a change in threshold) without a change in the slope (gain, sensitivity). This means that there is an additive interaction. The following important interpretation can be derived from Fig. 5A. At the heat gain side in a warm extrahypothalamic condition strong hypothalamic cooling has to be done to get a response. Under natural conditions this state is unlikely to occur. On the other hand, at cold extrahypothalamic temperatures the threshold for heat gain is shifted to high hypothalamic temperatures. This means that small variations of hypothalamic temperature in the normal range do affect the response; i.e., at peripheral cold load small variations in hypothalamic temperature can have an influence on the response magnitude and this is within the scope of natural conditions. On the heat loss side, similarly in a warm extrahypothalamic state, the threshold is at a low level and the response curve intersects with the normal temperature. Thus, small variations in hypothalamic temperature are followed by a change in response.

Figure 5B,C shows types of interaction where there is a change in the slope (sensitivity, gain). This type of interaction is a multiplicative one. Figure 5B depicts an increase in hypothalamic sensitivity on the heat gain or heat loss side in

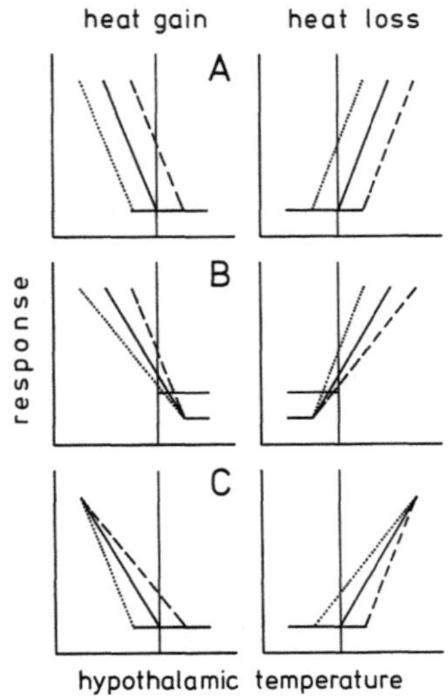

Fig. 5A – C. Types of thermoregulatory responses (heat loss, *right*; heat gain, *left*) to heating and cooling the hypothalamus and their variations at three different extrahypothalamic thermal conditions. **A** Change in threshold. **B, C** Change in slope (gain, sensitivity). *Solid line,* response at normal extrahypothalamic temperatures. *Dotted line,* response in warm extrahypothalamic state. *Dashed line,* response in cold extrahypothalamic state. (This figure applies, of course, to other combinations of stimulations.) *Vertical bar* shows normal value

cold or warm extrahypothalamic condition, respectively. The intersection point may lie below the baseline of the response (upper horizontal line). Figure 5C illustrates a decrease of hypothalamic thermosensitivity; e.g., in a cold extrahypothalamic state the heat gain response curve to hypothalamic stimulation is reduced in slope. At the heat loss side similarly at high extrahypothalamic temperatures the gain is reduced. In this type of response there is as clear a change in threshold as in Fig. 5A.

Another method of studying the interaction of different thermosensitive areas consists in assessing the response threshold in dependence on the two stimulation sites. This threshold may vary linearly (e.g., Rautenberg 1971) or follow a hyperbolic function (e.g., Brück and Wünnenberg 1970). The former points to an additive interaction (comparable to Fig. 5A) and the latter to a multiplicative interaction (comparable to Fig. 5B, C).

There is no principal difference in the interaction between skin (or ambient), total body core, spinal, and hypothalamic temperatures. Some recent investigations and the proposed types of interaction (as represented schematically in Fig. 5) will be cited here. They show the variety of responses also found in earlier investigations (Bligh 1973).

In a recent investigation of heat production in man an interaction between skin temperature and core temperature was found which followed Fig. 5B (Hayward et al. 1977). In a similar investigation (water bath) in rats the interaction between skin and core temperatures followed Fig. 5C with regard to heat production (Morhardt et al. 1975); this seems to be the only investigation to show this type of interaction on the heat gain side. In the goat the interaction between ambient temperature and core temperature with regard to heat production was not very prominent; no unequivocal type of interaction could be derived since curves at low and at high ambient temperature overlapped (Mercer and Jessen 1978). This probably resulted from the fact that the influence of skin temperature was low when compared with that of body core.

In recent investigations in rabbits (Stitt 1976; Boulant and Gonzalez 1977), the interaction between skin temperature and hypothalamic temperature followed

Fig. 5C with regard to heat loss and Fig. 5B with regard to heat gain (Boulant and Gonzalez 1977). In the goat the interaction of ambient temperature and spinal as well as hypothalamic temperature was studied with regard to heat gain and heat loss; both followed Fig. 5B (Jessen 1977).

In rats the interaction between ambient temperature and spinal temperature was proposed to follow Fig. 5A with regard to heat production (Banet and Hensel 1976). In pigeons both heat gain and heat loss threshold curves were of the shift in threshold type (Fig. 5A) as to the interaction between skin temperature and spinal temperature (Rautenberg 1971; Rautenberg et al. 1978). In penguins heat production followed the type depicted in Fig. 5B when combining ambient and spinal stimulation (Hammel et al. 1976).

Since there are usually large variations in the stimulus-response curves and since the intersection point may lie below the basic response shown in Fig. 5B or at extreme temperatures reflected in Fig. 5C, in a medium temperature range it seems to be often difficult to decide whether the curves at different extrahypothalamic temperatures are parallel or have a change in slope in one or another direction. In this way the distinction of different types of interaction is probably of little importance, especially when taking into account that, under natural conditions, large changes in hypothalamic (or generally central) temperatures never occur. In any case, a shift of the curves to the left is found at high extrahypothalamic temperatures, while a shift to the right is seen at low temperatures, and that is what is really significant.

3.4.2 Electrophysiological Evidence of the Interaction Between Different Thermosensitive Areas

The response of central neurons to combined stimulation of two thermosensitive areas is determined by the response to local stimulation – the hypothalamus may serve as an example in the following considerations – and by the response to the stimulation of the area projecting to the hypothalamus (or any other area). (The response to local stimulation is described in Sects. 3.1 – 3.3.) Before considering the response to combined stimulation the response to stimulation of the projecting area will be described; the response characteristics all follow the types schematically respresented in Fig. 4.

In most investigations the response to combined stimulation is studied by assessing the response curve to local (hypothalamic) stimulation in different thermal states (usually warm, normal, or cold) of the projecting area. Similar to the thermoregulatory response curves represented in Fig. 5, there should be a change in the response curves. Theoretically, the possible types of responses can be derived from a combination of all types shown in Fig. 4. In this way 36 types of interaction would be possible. But not all combinations occur and many do not make sense, e.g., an increase of activity in a warm-sensitive hypothalamic neuron, which is assumed to be the neurological basis for heat loss activation during spinal cooling, is nonhomeostatic and therefore inappropriate. (But one has to keep in mind that not all neurons responding to thermal stimulation must be involved in temperature regulation.)

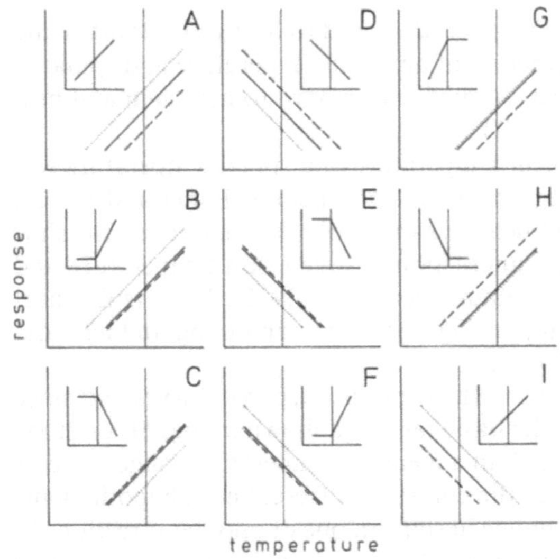

Fig. 6A – I. Variations of the response of hypothalamic neurons to local thermal stimulations in three different extrahypothalamic conditions (normal, *solid line*; cold, *dashed line*; warm, *dotted line*). *Inset figures,* response of the hypothalamic neuron to extrahypothalamic stimulation. The variations in the examples of **A – I** can be explained by combining the response to hypothalamic stimulation with the response to extrahypothalamic stimulation shown in the inset figures. *Vertical bar* shows normal value

Figure 6 shows selected response curves to local hypothalamic stimulation and its variation by stimulation of a projecting area, whose response characteristic is shown in the inset figure and which explains the shifts in the diagrams. The figure shall be explained by one example. Figure 6A shows the response of a hypothalamic warm-sensitive neuron with a linear characteristic (only linear hypothalamic characteristics are considered here for reasons of simplification; of course, interactions of threshold curves are also conceivable and do occur). Thermal stimulation of, e.g., the spinal cord may result in a similar response curve (linear, warm sensitive) of this hypothalamic neuron (see inset in Fig. 6A). From this it follows that at high spinal temperature the curve is shifted to the left (dotted line) as compared with the response curve at normal spinal temperature. At low spinal temperature the response curve is shifted to the right (dashed line). This type of response corresponds to Fig. 5, right side. Correspondingly, the other types depicted in Fig. 6 can be explained and reconstructed.

Interaction Between Skin and Spinal Cord. In the cat, spinal cold-sensitive neurons were often activated by skin cooling, the response being of the type shown in Fig. 4A (Simon 1972). This activation was found only in spinalized preparations whereas in intact animals no such influence from the skin was found (Bergheim-Hackmann and Simon 1976). The interaction between skin and spinal stimulation, i.e., a response of spinal thermosensitive neurons to skin stimulation, which was also found in pigeons (Necker 1975b), was not studied quantitatively. Nevertheless, usually cold-sensitive spinal neurons were activated by peripheral cooling and warm-sensitive neurons by peripheral warming.

Interaction Between Skin and Lower Brain Stem. Over 50% of the temperature-sensitive neurons in the rabbit's medulla responded to skin stimulation, but the influence was usually small (< 1 impulse/s °C in warm units and about −2 impulses/s °C in cold units). The responses were of the linear type (Fig. 4A). As to the interaction, sensitivity of the neurons to medullar stimulation was often opposite that to skin stimulation (Inoue and Murakami 1976).

Interaction Between Spinal Cord and Midbrain. In rabbits temperature-sensitive midbrain neurons responding to spinal thermal stimulation were found, and all responses were of the threshold type depicted in Fig. 4C (left side), i.e., spinal heating resulted in an inhibition (Hori and Harada 1976b). Altogether, only few temperature-sensitive neurons in the midbrain reticular formation responded to spinal stimulation. Since the temperature-sensitive neurons were usually cold sensitive and the response to spinal heating was of the warm-inhibited type, a combined response type as shown in Fig. 6E resulted (Hori and Harada 1976b).

Interaction of Skin or Spinal Cord and Hypothalamus. There are only few investigations of the response of POAH neurons to *skin stimulation,* especially with a quantitative assessment of the response curve which could be compared with the schematic responses presented in Fig. 4. In rabbits the response of POAH neurons to peripheral stimulation was generally of the threshold type, and all neurons had thresholds near the normal Tb (Boulant and Hardy 1974). All types of threshold responses shown in Fig. 4B and C were found. During selective stimulation of the rat's scrotum threshold neurons were again found (Nakayama et al. 1979), but thresholds varied over a wide temperature range. This might be a specialization of the scrotal system (see Sect. 1.2).

In a detailed study in rabbits (Guieu and Hardy 1970), POAH neurons responded to *spinal stimulation* and the response was generally of the threshold type with thresholds near Tb. All four types represented in Fig. 4B and C were found. In another investigation in rabbits (Boulant and Hardy 1974), in addition to the threshold type, linear responses were also found (Fig. 4A). In a study in pigeons there was again a predominance of the threshold type and a preponderance of the type shown in Fig. 4B (right side), but the other types were also found and cold units were generally of the type depicted in Fig. 4A (Rosner 1978).

The interaction between *skin and hypothalamus* was studied systematically in rabbits and cats (Hellon 1972a), but only a few neurons were found which were activated both by preoptic and skin stimulation. Two neurons, linear and warm-sensitive with regard to preoptic thermal stimulation, showed a shift to the left during peripheral cooling, i.e., the response to skin stimulation was opposite that of preoptic stimulation (see Fig. 6H). This response does not fit with the response models shown in Fig. 5. In one unit (linear, warm sensitive) skin cooling inhibited the response to preoptic stimulation (Fig. 6G).

In investigations in ground squirrels (Boulant and Bignall 1973b) and in rabbits (Boulant and Hardy 1974), most neurons had the same thermal coefficient to local hypothalamic and peripheral or spinal stimulation. As to the interaction between skin or spinal cord and hypothalamus, different types were found. Most interactions were of the type shown in Fig. 6A. There was a decrease in sensitivity (change in gain or slope) at high extrahypothalamic temperatures, which corresponds to the thermoregulatory type illustrated in Fig. 5C. This agrees with the results of thermoregulatory responses (Boulant and Gonzalez 1977; Stitt 1976). Other interactions were of the types shown in Fig. 6B, C, E, F, and I. Of these types, 6C, F, and I are inappropriate (i.e., they do not fit with the response types depicted in Fig. 5).

In a hibernator (golden hamster) hypothalamic thermosensitivity was changed by skin temperature (Speulda and Wünnenberg 1977). The type of interaction was principally that of Fig. 6A, but at very low preoptic temperatures it changed to that of Fig. 6H, i.e., activity was higher at low skin temperature than at high skin temperatures. In this way the sensitivity was extended to lower temperatures, which is clearly important in the hibernating state where both central and peripheral temperatures are low.

3.5 Hypothalamic Thermosensitive Neurons Under Various Conditions

3.5.1 Firing Pattern

The criterion generally used to classify a neuron as thermosensitive is a change in the *mean firing rate*. It has been shown that local hypothalamic thermal stimulation can alter the firing pattern of some neurons without altering the mean frequency (Jahns and Werner 1974; Reaves and Heath 1975). It is not clear whether such responses play an important role in thermoregulation, but the number of responding cells increases when including such responses.

3.5.2 Hypothalamic Tissue Cultures

Temperature-sensitive neurons have been demonstrated in vivo in tissue cultures. Some of the hypothalamic neurons were active only within a narrow temperature range near 36° – 37 °C (Mason et al. 1978). In POAH tissue cultures of mice both warm-sensitive and cold-sensitive neurons were found, and the proportion of warm-sensitive units was similar to that observed in in vivo experiments (Nakayama et al. 1978).

3.5.3 Effects of Pyrogen and Acetylsalicylate

Shortly after the discovery of hypothalamic temperature-sensitive neurons their response to pyrogen was tested to see whether pyrogen may act by influencing hypothalamic temperature-sensitive neurons. It was shown that pyrogen decreased the activity and sensitivity of warm-sensitive units and increased activity in cold-sensitive neurons; temperature-insensitive neurons were not affected. The antipyretic agent acetylsalicylate depressed the effect of pyrogen (Hellon 1974). From this it was concluded that the increase of Tb during fever could result from a reduced sensitivity of warm-sensitive neurons, which reduces heat loss and an increase in heat production by the increased activity in cold-sensitive neurons.

Recently the effect of pyrogen on temperature-sensitive neurons was studied in other brain areas. In the midbrain reticular formation, where cold-sensitive neurons dominate, all such neurons increased activity after pyrogen injection (Nakayama and Hori 1973). In the medulla warm-sensitive neurons were depressed and cold-sensitive neurons excited by pyrogen (Sakata 1979); an antipyretic agent abolished the effect of pyrogen. From these results it may be deduced that pyrogen and antipyretic substances act in a uniform manner on temperature-sensitive neurons (nonsensitive neurons were generally not affected), perhaps by interacting with the temperature-sensitive processes at the neuronal membrane (cf. the opposite effect of Ca^{++} on cold- and warm-receptors!).

3.6 Thermosensitivity of the CNS: Specific or Nonspecific Response?

The question arises whether there are really temperature-sensitive primary sensory neurons in the CNS comparable to the peripheral thermoreceptors or whether thermal stimulation influences the activity of interneurons in the thermoregulatory system. These two possibilities are represented schematically in Fig. 7. The upper model shows the case of receptor neurons in the CNS and the lower model the case of a nonspecific effect of temperature on the activity of neurons involved in afferent and efferent neuronal circuits of the thermoregulatory system (the site of thermal sensitivity is marked by a circle).

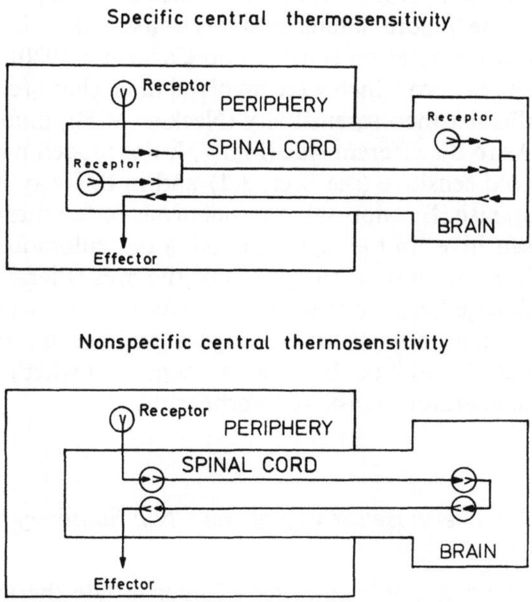

Fig. 7. Models of CNS thermosensitivity. In the *upper model* specific thermoreceptors are assumed to exist both in the periphery and in the CNS. In the *lower model* thermoreceptors are assumed to exist only in the periphery. Thermal stimulation of the CNS in this case is assumed to influence the activity of interneurons of the thermoregulatory system. A combination of both models is also conceivable. *Circles* indicate the site of thermal sensitivity

According to Eisenman (1972) a characteristic feature of sensory neurons is a linear response to thermal stimulation over a wide temperature range (see. Fig. 4A). These neurons were found to be less sensitive to anesthetics than the threshold neurons (Eisenman 1972). On the other hand, such sensory neurons should not have an input from other thermosensitive areas. Actually, part of the linear neurons was found to have such an input (see Sect. 3.4). In addition, linear neurons in the hypothalamus were found to receive an input from the brain stem as shown by electrical stimulation (Eisenman 1974). From this the existence of true sensory neurons is not supported.

In the nonspecific model depicted in Fig. 7 there are specific receptors only in the periphery. Thermal stimulations of central areas are assumed to influence the transmission of signals from one cell to another or the activity of the interneuron both at the afferent and at the efferent side of the thermoregulatory system; axonal conduction is unlikely to be influenced (Paintal 1965). The results described in Sect. 3.4, where neurons which responded to skin stimulation were also influenced by local central stimulation in many cases, point to a nonspecific effect. To get an appropriate thermoregulatory response one must expect that more interneurons of the warm receptor afferent pathway are warm sensitive than cold sensitive, and vice versa, for the cold receptor afferent pathway to get a cooperative response to central stimulation. This was generally found in electrophysiological experiments (see Sect. 3.4).

Although this has not been studied systematically there is some evidence that not only neurons of the thermoregulatory system are thermosensitive. In the cortex of cats neurons were found which had a similar thermosensitivity as those found in the hypothalamus, and it is unlikely that these neurons are involved in temperature regulation (Barker and Carpenter 1970). Many dorsal horn neurons in the spinal cord which were involved in mechanoreception or pain reception showed a distinct thermosensitivity (Necker 1975b; unpublished observations).

As to the efferent side (Fig. 7, lower model) motoneurons have been shown to be cold sensitive (see Sect. 3.1) and in this way may cooperate in the shivering response. But not only motoneurons in the thermoregulatory systems are thermosensitive. In the pigeon, cooling of oculomotor nuclei in the brain stem resulted in a response of the pupils of the eyes (Pierau et al. 1970).

Altogether, the thermosensitivity of neurons probably involved in temperature regulation seems to be cooperative in such a manner that appropriate responses occur, and it might be not so important whether there are true sensory neurons or temperature-sensitive interneurons.

3.7 Thermosensitivity of the CNS: Summary and Conclusion

There is ample evidence of appropriate thermoregulatory responses to thermal stimulation of the CNS. That there is generally a static deviation of Tb during central stimulation may result from the stimulation of a selected area, a situation which does not occur under natural conditions; a similar deviation was also found during selective stimulation of a small skin area (e.g., scrotal heating in the ram; Waites 1962).

Since the response must be based on a change in the activity of central neurons it is not surprising that, in areas where thermal stimulation elicited thermoregulatory responses, temperature-sensitive neurons were found. Since thermoregulatory responses are characterized by a threshold, neurons were found with threshold characteristics and thresholds were often near normal Tb. The shift in response during combined stimulations of different thermosensitive areas was also found in electrophysiological investigations. Response patterns were found which fit with the thermoregulatory responses, but the variety of patterns also included responses which do not make sense.

Altogether, the demonstration of temperature-sensitive neurons does not prove that the neurons recorded from are actually involved in temperature regulation, since neurons which clearly do not belong to the thermoregulatory system have also been shown to be temperature sensitive.

Besides homeostatic responses there is accumulating evidence of nonhomeostatic reactions to thermal stimulation of the CNS. In birds preoptic hypothalamic cooling and in goats posterior hypothalamic temperature changes resulted in nonhomeostatic responses. In addition, it has been shown that thermal stimulation of the hypothalamus also influences other autonomic systems, e.g. osmoregulation (Simon-Oppermann and Jessen 1977; Simon-Oppermann et al. 1979).

A final point concerns the correlation between natural variations of CNS temperature and thermoregulatory activity. The responses shown in selective thermal stimulations indicate that, at corresponding natural variations under certain conditions, a contribution of the thermosensitivity of the CNS to the overall thermoregulatory activity should occur. But this contribution of natural variations has not yet been demonstrated convincingly.

References

Abrams R, Hammel HT (1964) Hypothalamic temperature in unanesthetized albino rats during feeding and sleeping. Am J Physiol 206:641 – 646

Abrams R, Hammel HT (1965) Cyclic variations in hypothalamic temperature in unanesthetized rats. Am J Physiol 208:698 – 702

Aschoff J (1970) Circadian rhythm of activity and of body temperature. In: Hardy JD, Gagge AP, Stolwijk JAJ (eds) Physiological and behavioral temperature regulation. Thomas, Springfield, pp 905 – 919

Bade H, Braun HA, Hensel H (1979) Parameters of the static burst discharge of lingual cold receptors in the cat. Pfluegers Arch 382:1 – 5

Banet M, Hensel H (1976) The interaction between cutaneous and spinal thermal inputs in the control of oxygen consumption in the rat. J Physiol (Lond) 260:461 – 473

Barker JL, Carpenter DO (1970) Thermosensitivity of neurons in the sensorimotor cortex of the cat. Science 169:597 – 598

Benzinger TH (1970) Peripheral cold reception and central warm reception, sensory mechanisms of behavioral and autonomic thermostasis. In: Hardy JD, Gagge AP, Stolwijk JAJ (eds) Physiological and behavioral temperature regulation. Thomas, Springfield, pp 831 – 855

Bergheim-Hackmann E, Simon E (1976) Der Einfluß von Haut- und Rückenmarkstemperaturen auf aufsteigende temperaturempfindliche Fasern des Rückenmarks bei der Hauskatze (Felis domestica Briss). Verh Dtsch Zool Ges 278

Bernstein MH (1974) Vascular responses and foot temperature in pigeons. Am J Physiol 226:1350–1355
Bligh J (1972) Neuronal models of mammalian temperature regulation. In: Bligh J, Moore RE (eds) Essays on temperature regulation. North-Holland, Amsterdam London, pp 105–120
Bligh J (1973) Temperature regulation in mammals and other vertebrates. North-Holland, Amsterdam
Bligh J (1979) The central neurology of mammalian thermoregulation. Neuroscience 4:1213–1236
Boivie J (1979) An anatomical reinvestigation of the termination of the spinothalamic tract in the monkey. J Comp Neurol 186:343–370
Boulant JA (1974) The effect of firing rate on preoptic neuronal thermosensitivity. J Physiol (Lond) 240:661–669
Boulant JA, Bignall KE (1973a) Determinants of hypothalamic neuronal thermosensitivity in ground squirrels and rats. Am J Physiol 225:306–310
Boulant JA, Bignall KE (1973b) Hypothalamic neuronal responses to peripheral and deep-body temperatures. Am J Physiol 225:1371–1374
Boulant JA, Gonzalez RR (1977) The effect of skin temperature on the hypothalamic control of heat loss and heat production. Brain Res 120:367–372
Boulant JA, Hardy JD (1974) The effect of spinal and skin temperatures on the firing rate and thermosensitivity of preoptic neurones. J Physiol (Lond) 240:639–660
Brooks CM, Koizumi K, Malcolm JL (1955) Effects of changes in temperature on reactions of spinal cord. J Neurophysiol 18:205–216
Brück K, Wünnenberg W (1970) "Meshed" control of two effector systems: nonshivering and shivering thermogenesis. In: Hardy JD, Gagge AP, Stolwijk JAJ (eds) Physiological and behavioral temperature regulation. Thomas, Springfield, pp 562–580
Bullard RW, Banerjee MR, Chen F, Elizondo R, MacIntyre BA (1970) Skin temperature and thermoregulatory sweating: a control systems approach. In: Hardy JD, Gagge AP, Stolwijk JAJ (eds) Physiological and behavioral temperature regulation. Thomas, Springfield, pp 597–610
Burton H (1975) Responses of spinal cord neurones to systematic changes in hind limb skin temperatures in cats and primates. J Neurophysiol 38:1060–1079
Burton H, Forbes DJ, Benjamin RM (1970) Thalamic neurons responsive to temperature changes of glabrous hand and foot skin in squirrel monkey. Brain Res 24:179–190
Cabanac M (1975) Temperature regulation. Ann Rev Physiol 37:415–439
Cabanac M (1979) Le comportement thermorégulateur. J Physiol (Paris) 75:115–178
Cabanac M, Hardy JD (1969) Responses unitaires et thermorégulatrices lors de réchauffement et refroidissement localisés de la région préoptique et du mésencephale chez le lapin. J Physiol (Paris) 61:331–347
Carlisle HJ, Ingram DI (1973) The influence of body core temperature and peripheral temperatures on oxygen consumption in the pig. J Physiol (Lond) 231:341–352
Carpenter DO (1970) Membrane potential produced directly by the Na^+-pump in aplysia neurons. Comp Biochem Physiol 75:371–385
Chai CY, Lin MT (1972) Effects of heating and cooling the spinal cord and medulla oblongata on thermoregulation in monkeys. J Physiol (Lond) 225:297–308
Chai CY, Lin MT (1973) Effects of thermal stimulation of medulla oblongata and spinal cord on decerebrate rabbits. J Physiol (Lond) 234:409–419
Chambers MR, Andres KH, Düring M von, Iggo A (1972) The structure and function of the slowly adapting type II mechanoreceptor in hairy skin. Q J Exp Physiol 57:417–445
Christensen BN, Perl ER (1970) Spinal neurons specifically excited by noxious or thermal stimuli: marginal zone of the dorsal horn. J Neurophysiol 33:293–307
Clark WG (1979) Changes in body temperature after administration of amino acids, peptides, dopamine, neuroleptics and related agents. Neurosci Biobehav Rev. 3:179–231

Corbit JD (1970) Behavioral regulation of body temperature. In: Hardy JD, Gagge AP, Stolwijk JAJ (eds) Physiological and behavioral temperature regulation. Thomas, Springfield, pp 777–801
Cranston WI, Hellon RF, Townsend Y (1977) Are there functionally important temperature sensors in the right heart or lungs? J Physiol (Lond) 273:533–537
Cranston WI, Hellon RF, Townsend Y (1978) Thermal stimulation of intra-abdominal veins in conscious rabbits. J Physiol (Lond) 277:49–52
Crawshaw LI, Nadel ER, Stolwijk JAJ, Stamford BA (1975) Effect of local cooling on sweating rate and cold sensation. Pfluegers Arch 354:19–28
Cronin MJ, Baker MA (1976) Heat-sensitive midbrain raphe neurons in the anesthetized cat. Brain Res 110:175–181
Cronin MJ, Baker MA (1977a) Thermosensitive midbrain neurons in the cat. Brain Res 128:461–472
Cronin MJ, Baker MA (1977b) Physiological responses to midbrain thermal stimulation in the cat. Brain Res 128:542–546
Darian-Smith I, Johnson KO, Dykes R (1973) "Cold" fiber population innervating palmar and digital skin of the monkey: responses to cooling pulses. J Neurophysiol 36:325–346
Darian-Smith I, Johnson KO, LaMotte C, Kenins P, Shigenaga Y, Ming VC (1979) Coding of incremental changes in skin temperature by single warm fibers in the monkey. J Neurophysiol 42:1316–1331
Dawson WR, Hudson JW (1970) Birds. In: Whittow GC (ed) Comparative physiology of thermoregulation, vol I. Academic Press, New York London, pp 223–310
Dickenson AH (1977) Specific responses of rat raphe neurones to skin temperature. J Physiol (Lond) 273:277–293
Dickenson AH, Hellon RF, Taylor DCM (1979) Facial thermal input to the trigeminal spinal nucleus of rabbits and rats. J Comp Neurol 185:203–210
Dodt E (1952) The behaviour of thermoreceptors at low and high temperatures with special reference to Ebbecke's temperature phenomena. Acta Physiol Scand 27:296–314
Dostrovsky JO, Hellon RF (1978) The representation of facial temperature in the caudal trigeminal nucleus of the cat. J Physiol (Lond) 277:29–47
Dubner R, Sumino R, Wood WJ (1975) A peripheral "cold" fiber population responsive to innocuous and noxious thermal stimuli applied to monkey's face. J Neurophysiol 38:1373–1389
Duclaux R (1977) Les récepteurs thermique cutanées. J Physiol (Paris) 73:849–862
Duclaux R, Kenshalo DR (1972) The temperature sensitivity of the type I slowly adapting mechanoreceptors in cats and monkeys. J Physiol (Lond) 224:647–664
Duclaux R, Kenshalo DR (1973) Cutaneous receptive fields of primate cold fibers. Brain Res 55:437–442
Duclaux R, Kenshalo DR, Sr (1980) Response characteristics of cutaneous warm receptors in the monkey. J Neurophysiol 43:1–15
Dykes RW (1975) Coding of steady and transient temperatures by cutaneous 'cold' fibers serving the hand of monkeys. Brain Res 98:485–500
Edinger HM, Eisenman JS (1970) Thermosensitive neurons in tuberal and posterior hypothalamus of cats. Am J Physiol 219:1098–1103
Eidelberg E, Rick C (1975) Lack of effect of partial spinal cord sections upon thermal discrimination in the monkey. Appl Neurophysiol 38:145–152
Eisenman JS (1972) Unit activity studies of thermoresponsive neurons. In: Bligh J, Moore RE (eds) Essays on temperature regulation. North Holland, Amsterdam London, pp 55–69
Eisenman JS (1974) Unit studies of brainstem projection to the preoptic area and hypothalamus. In: Lederis K, Cooper KE (eds) Recent studies of hypothalamic function. Karger, Basel, pp 328–340
Emmers R (1966) Separate relays of tactile, pressure, thermal and gustatory modalities in the cat thalamus. Proc Soc Exp Biol Med 121:527–531

Fruhstorfer H (1976) Conduction in the afferent thermal pathways of man. In: Zotterman Y (ed) Sensory functions of the skin in primates. Pergamon, Oxford New York Toronto, pp 355 – 366
Fruhstorfer H, Hensel H (1973) Thermal cutaneous afferents in the trigeminal nucleus of the cat. Naturwissenschaften 60:209
Fruhstorfer H. Zenz M, Nolte H, Hensel H (1974) Dissociated loss of cold and warm sensibility during regional anaesthesia. Pfluegers Arch 349:73 – 82
Gallego R, Eyzaguirre C, Monti-Bloch L (1979) Thermal and osmotic responses of arterial receptors. J Neurophysiol 42:665 – 680
Görke K (1980) Influences of spinal cord temperature changes on reflex discharge and spontaneous activity of spinal motoneurones in pigeons and leguans. J Comp Physiol 139:251 – 259
Görke K, Pierau F-K (1979) Initiation of muscle activity in spinalized pigeons during spinal cord cooling and warming. Pfluegers Arch 381:47 – 52
Görke K, Necker R, Rautenberg W (1975) Neurophysiological investigation of spinal reflexes at different temperatures of the spinal cord in birds and reptiles. Pfluegers Arch 359:269 – 271
Gorman ALF, Marmor MF (1970) Temperature dependence of the sodium-potassium permeability ratio of a molluscan neurone. J Physiol (Lond) 210:919 – 931
Graf R (1980a) Diurnal changes of thermoregulatory functions in pigeons: I. Effector mechanisms. Pfluegers Arch 386:173 – 179
Graf R (1980b) Diurnal changes of thermoregulatory functions in pigeons: II. Spinal thermosensitivity. Pfluegers Arch 386:181 – 185
Graf R, Necker R (1979) Cyclic and non-cyclic variations of spinal cord temperature related with temperature regulation in pigeons. Pfluegers Arch 380:215 – 220
Gregory JF (1973) An electrophysiological investigation of the receptor apparatus of the duck's bill. J Physiol (Lond) 229:151 – 164
Guieu JD, Hardy JD (1970) Effects of heating and cooling of the spinal cord on preoptic unit activity. J Appl Physiol 29:675 – 683
Gupta BN, Nier K, Hensel H (1979) Cold-sensitive afferents from the abdomen. Pfluegers Arch 380:203 – 204
Hainsworth FR, Stricker EM (1970) Salivary cooling by rats in the heat. In: Hardy JD, Gagge AP, Stolwijk JAJ (eds) Physiological and behavioral temperature regulation. Thomas, Springfield, pp 611 – 626
Hales JRS, Hutchison JCD (1971) Metabolic respiratory and vasomotor responses to heating the scrotum of the ram. J Physiol (Lond) 212:353 – 375
Hammel HT (1972) The set-point in temperature regulation: analogy or reality. In: Bligh J, Moore RE (eds) Essays on temperature regulation. North-Holland, Amsterdam London, pp 121 – 137
Hammel HT, Maggert J, Kaul R, Simon E, Simon-Oppermann C (1976) Effects of altering spinal cord temperature on temperature regulation in the Adelie penguins. Pfluegers Arch 362:1 – 6
Hardy JD (1969) Thermoregulatory responses to temperature changes in the midbrain of the rabbit. Fed Proc 28:713
Hardy JD (1973) Posterior hypothalamus and the regulation of body temperature. Fed Proc 32:1564 – 1571
Hayward JS, Eckerson JD, Collis ML (1977) Thermoregulatory heat production in man: prediction equation based on skin and core temperatures. J Appl Physiol 42:377 – 384
Heller HC (1979) Hibernation – neural aspects. Ann Rev Physiol 41:305 – 321
Heller HC, Glotzbach SF (1977) Thermoregulation during sleep and hibernation. Int Rev Physiol 15:147 – 188
Heller HC, Walker JM, Forant SL, Glotzbach SF, Berger RJ (1978) Sleep and hibernation: electrophysiological and thermoregulatory homologies. In: Wang L, Hudsor J

(eds) Strategies in cold: Natural torpidity and thermogenesis. Academic Press, New York, pp 225–265
Hellon RF (1967) Thermal stimulation of hypothalamic neurones in unanaesthetized rabbits. J Physiol (Lond) 193:381–395
Hellon RF (1972a) Temperature-sensitive neurons in the brain stem: Their responses to brain temperature at different ambient temperatures. Pfluegers Arch 335:323–334
Hellon RF (1972b) Central thermoreceptors and thermoregulation. In: Neil E (ed) Enteroreceptors. Springer, Berlin Heidelberg New York (Handbook of sensory physiology, vol III/1, pp 161–186)
Hellon RF (1974) Monoamines, pyrogens and cations: Their actions on central control of body temperature. Pharmacol Rev 26:289–321
Hellon RF, Misra NK (1973a) Neurones in the dorsal horn of the rat responding to scrotal skin temperature changes. J Physiol (Lond) 232:375–388
Hellon RF, Misra NK (1973b) Neurones in the ventrobasal complex of the rat thalamus responding to scrotal skin temperature changes. J Physiol (Lond) 232:389–400
Hellon RF, Mitchell D (1975) Convergence in a thermal afferent pathway in the rat. J Physiol (Lond) 248:359–376
Hellon RF, Misra NK, Provins KA (1973) Neurones in the somatosensory cortex of the rat responding to scrotal skin temperature changes. J Physiol (Lond) 232:401–411
Hellon RF, Hensel H, Schäfer K (1975) Thermal receptors in the scrotum of the rat. J Physiol (Lond) 248:349–357
Hensel H (1973a) Neural processes in thermoregulation. Physiol Rev 53:948–1017
Hensel H (1973b) Cutaneous thermoreceptors. In: Iggo A (ed) Somatosensory system. Springer, Berlin Heidelberg New York (Handbook of sensory physiology, vol II, pp 79–110)
Hensel H (1974) Thermoreceptors. Annu Rev Physiol 36:233–249
Hensel H (1976) Functional and structural basis of thermoreception. Prog Brain Res 43:105–118
Hensel H, Banet M (1978) Thermoreceptor activity, insulative and metabolic changes in cold and warm adapted cats. In: Houdas Y, Guieu JD (eds) New trends in thermal physiology. Masson, Paris New York Barcelona, pp 53–55
Hensel H, Boman KKA (1960) Afferent impulses in cutaneous sensory nerves in human subjects. J Neurophysiol 23:564–578
Hensel H, Iggo A (1971) Analysis of cutaneous warm and cold fibers in primates. Pfluegers Arch 329:1–8
Hensel H, Nier K (1976) Total shift from cold to warm sensitivity of the ampullae of Lorenzini. Naturwissenschaften 63:147
Hensel H, Schäfer K (1974) Effects of calcium on warm and cold receptors. Pfluegers Arch 352:87–90
Hensel H, Schäfer K (1979) Activity of cold receptors in cats after long-term adaptation to various temperatures. Pfluegers Arch [Suppl] 379:R56
Hensel H, Wurster RD (1970) Static properties of cold receptors in nasal area of cats. J Neurophysiol 33:271–275
Hensel H, Brück K, Raths P (1973) Homeothermic organisms. In: Precht H, Christophersen J, Hensel H, Larcher W (eds) Temperature and life. Springer, Berlin Heidelberg New York, pp 503–761
Hensel H, Andres KH, von Düring M (1974) Structure and function of cold receptors. Pfluegers Arch 352:1–10
Herdman SJ (1978) Recovery of shivering in spinal cats. Exp. Neurol 59:177–189
Hertel H-C, Howaldt B, Mense S (1976) Responses of group IV and group III muscle afferents to thermal stimuli. Brain Res 113:201–205
Hori T, Harada Y (1976a) Responses of midbrain raphe neurons to local temperature. Pfluegers Arch 364:205–207

Hori T, Harada Y (1976b) Midbrain neuronal responses to local and spinal cord temperatures. Am J Physiol 231:1573 – 1578
Iggo A (1969) Cutaneous thermoreceptors in primates and subprimates. J Physiol (Lond) 200:403 – 430
Iggo A, Paintal AS (1977) The metabolic dependence of primate cutaneous cold receptors. J Physiol (Lond) 272:40 – 41
Iggo A, Ramsey RL (1974) Dorsal horn neurones excited by cutaneous cold receptors in primates. J Physiol (Lond) 242:132 – 133
Iggo A, Ramsey RL (1976) Thermosensory mechanisms in the spinal cord of monkeys. In: Zotterman Y (ed) Sensory functions of the skin in primates. Pergamon, Oxford New York Toronto, pp 285 – 304
Ingram DL, Legge KF (1972) The influence of deep body and skin temperatures on thermoregulatory responses to heating of the scrotum in pigs. J Physiol (Lond) 224:477 – 487
Inoue S, Murakami N (1976) Unit responses in the medulla oblongata of rabbit to changes in local and cutaneous temperature. J Physiol (Lond) 259:339 – 356
Jahns R (1975) Types of neuronal responses in the rat thalamus to peripheral temperature changes. Exp Brain Res 23:157 – 166
Jahns R (1976) Different projections of cutaneous thermal inputs to single units of the midbrain raphe nuclei. Brain Res 101:355 – 361
Jahns R, Werner J (1974) Analysis of periodic components of hypothalamic spike – trains after central thermal stimulation. Pfluegers Arch 351:13 – 24
Jessen C (1977) Interaction of air temperature and core temperatures in thermoregulation of the goat. J Physiol (Lond) 264:585 – 606
Jessen, C, Simon-Oppermann C (1976) Production of temperature signals in the peripherally denervated spinal cord of the dog. Experientia 32:484 – 485
Johansen K, Millard RW (1973) Vascular responses to temperature in the foot of the giant fulmar, Macronectes giganteus. J Comp Physiol 85:47 – 64
Johnson JM, Park MK (1979) Reflex control of skin blood flow by skin temperature: role of core temperature. J Appl Physiol 47:1188 – 1193
Keatinge WR (1970) Direct effects of temperature on blood vessels: their role in cold vasodilatation. In: Hardy JD, Gagge AP, Stolwijk JAJ (eds) Physiological and behavioral temperature regulation. Thomas, Springfield, pp 231 – 236
Kenshalo DR, Duclaux R (1977) Response characteristics of cutaneous cold receptors in the monkey. J Neurophysiol 40:319 – 332
Kenshalo DR, Cormier D, Mellos M (1976) Some response properties of cold fibers to cooling. Prog Brain Res 43:129 – 142
Kluger MJ, Gonzalez RP, Hardy JD (1972) Peripheral thermal sensitivity in the rabbit. Am J Physiol 222:1031 – 1034
Klussmann FW, Pierau F-K (1972) Extrahypothalamic deep body thermosensitivity. In: Bligh J, Moore RE (eds) Essays on temperature regulation. North-Holland, Amsterdam London, pp 87 – 104
Knox, GV, Campbell C, Lomax P (1973) Cutaneous temperature and unit activity in the hypothalamic thermoregulatory centers of the rat. Exp Neurol 40:717 – 730
Konietzny F, Hensel H (1975) Warm fiber activity in human skin nerves. Pfluegers Arch 359:265 – 267
Konietzny F, Hensel H (1977) The dynamic response of warm units in human skin nerves. Pfluegers Arch 370:111 – 114
Kosaka M, Simon E (1968) Kältetremor wacher chronisch spinalisierter Kaninchen im Vergleich zum Kältezittern intakter Tiere. Pfluegers Arch 302:333 – 356
Kreisman NR, Zimmermann ID (1973) Representation of information about skin temperature in the discharge of single cortical neurons. Brain Res 55:343 – 354
Kumazawa T, Perl ER (1977) Primate cutaneous receptors with unmyelinated (C) fibres and their projection to the substantia gelatinosa. J Physiol (Paris) 73:287 – 304

Kumazawa T, Perl ER (1978) Excitation of marginal and substantia gelatinosa neurons in the primate spinal cord: indications of their place in dorsal horn functional organization. J Comp Neurol 177:417–434
Kumazawa T, Perl ER, Burgess PR, Whitehorn D (1975) Ascending projections from marginal zone (lamina I) neurons of the spinal dorsal horn. J Comp Neurol 162:1–12
LaMotte RH, Campbell JN (1978) Comparison of responses of warm and nociceptive C-fiber afferents in monkey with human judgments of thermal pain. J Neurophysiol 41:509–528
Landgren S (1957) Cortical reception of cold impulses from the tongue of the cat. Acta Physiol Scand 40:202–209
Landgren S (1960) Thalamic neurones responding to cooling of the cat's tongue. Acta Physiol Scand 48:255–267
Leitner LM, Roumy M (1974) Thermosensitive units in the tongue and in the skin of the duck's bill. Pfluegers Arch 346:151–156
Lipton JM (1973) Thermosensitivity of medulla oblongata in control of body temperature. Am J Physiol 224:890–897
Long RR (1977) Sensitivity of cutaneous cold fibers to noxious heat: paradoxical cold discharge. J Neurophysiol 40:489–502
Lynch WC, Adair ER, Adams PW (1980) Vasomotor thresholds in the squirrel monkey: effects of central and peripheral temperature. J Appl Physiol 48:89–96
Martin HF, Manning JW (1971) Thalamic 'warming' and 'cooling' units responding to cutaneous stimulation. Brain Res 27:377–381
Mason P, Hasan H, Valis M (1978) Spontaneous firing of hypothalamic neurones over a narrow temperature interval. Nature 273:242–243
McCaffrey TV, Wurster RD, Jacobs HK, Euler DE, Geis GS (1979) Role of skin temperature in the control of sweating. J Appl Physiol 47:591–597
McCook RD, Randall WC, Hassler CR, Mihaldzic N, Wurster RD (1970) The role of cutaneous thermal receptors in sudomotor control. In: Hardy JD, Gagge AP, Stolwijk JAJ (eds) Physiological and behavioral temperature regulation. Thomas, Springfield, pp 627–633
Mense S (1978) Effects of temperature on the discharges of muscle spindles and tendon organs. Pfluegers Arch 374:159–166
Mercer JB, Jessen C (1978) Central thermosensitivity in conscious goats: hypothalamus and spinal cord versus residual inner body. Pfluegers Arch 374:179–186
Mercer JB, Jessen C, Pierau F-K (1978) Thermal stimulation of neurons in the rostral brain stem of conscious goats. J Thermal Biol 3:5–10
Meurer KA, Jessen C, Iriki M (1967) Kältezittern während isolierter Kühlung des Rückenmarks nach Durchschneidung der Hinterwurzeln. Pfluegers Arch 293:236–255
Mills SH, Heath JE (1972) Responses to thermal stimulation of the preoptic area in the house sparrow, Passer domesticus. Am J Physiol 222:914–919
Molinari HH, Kenshalo DR (1977) Effect of cooling rate on the dynamic response of cat cold units. Exp Neurol 55:546–555
Morhardt JE, Lattanzi D, Miller C (1975) Metabolic responses of unanesthetized rats to manipulation of skin temperature. Am J Physiol 228:575–580
Mosso JA, Kruger L (1973) Receptor categories represented in spinal trigeminal nucleus caudalis. J Neurophysiol 36:472–488
Nadel ER, Bullard RW, Stolwijk JAJ (1971) Importance of skin temperature in the regulation of sweating. J Appl Physiol 31:80–87
Nadel ER, Mitchell JW, Stolwijk JAJ (1973) Differential thermal sensitivity in the human skin. Pfluegers Arch 340:71–76
Nakayama T, Hori T (1973) Effects of anesthetic and pyrogen on thermally sensitive neurons in the brainstem. J Appl Physiol 34:351–355
Nakayama T, Hori Y, Suzuki M, Yonezawa T, Yamamoto K (1978) Thermo-sensitive neurons in preoptic and anterior hypothalamic tissue cultured in vitro. Neurosci Lett 9:23–26

Nakayama T, Ishikawa Y, Tsurutani T (1979) Projection of scrotal thermal afferents to the preoptic and hypothalamic neurons in rats. Pfluegers Arch 380:59 – 64
Necker R (1972) Response of trigeminal ganglion neurons to thermal stimulation of the beak in pigeons. J Comp Physiol 78:307 – 314
Necker R (1973) Temperature-sensitivity of thermoreceptors and mechanoreceptors on the beak of pigeons. J Comp Physiol 87:379 – 391
Necker R (1975a) Temperature-sensitive ascending neurons in the spinal cord of pigeons. Pfluegers Arch 353:275 – 286
Necker R (1975b) Temperature sensitivity of spinal cord in pigeons: an electrophysiological investigation. In: Jansky L (ed) Depressed metabolism and cold thermogenesis. Charles University, Prague, pp 202 – 206
Necker R (1977) Thermal sensitivity of different skin areas in pigeons. J Comp Physiol 116:239 – 246
Necker R, Rautenberg W (1975) Effect of spinal deafferentation on temperature regulation and spinal thermosensitivity in pigeons. Pfluegers Arch 360:287 – 299
Necker R, Reiner B (1980) Temperature-sensitivity mechanoreceptors, thermoreceptors and heat nociceptors in the feathered skin of pigeons. J Comp Physiol 135:201 – 207
Neya T, Pierau F-K (1976) Vasomotor response to thermal stimulation of the scrotal skin in rats. Pfluegers Arch 363:15 – 18
Nieuwenhuys R, Voogd J, van Huijzen C (1978) The human central nervous system. A synopsis and atlas. Springer, Berlin Heidelberg New York
Norrsell U (1979) Thermosensory defects after cervical spinal cord lesions in the cat. Exp Brain Res 35:479 – 494
Norrsell U, Ullman M (1978) Note on the conduction velocity of warm afferent fibres from the skin of the human leg. Acta Physiol Scand 103:337 – 339
Paintal AS (1965) Effects of temperature on conduction in single vagal and saphenous myelinated nerve fibres of the cat. J Physiol (Lond) 180:20 – 49
Pierau F-K, Wurster RD (1975) Effects of ouabain and calcium on temperature responses of the cat tongue. Pfluegers Arch 359:R97
Pierau F-K, Alexandridis E, Spaan G, Oksche A, Klussmann FW (1976) Der Einfluß von lokalen Temperaturänderungen im pupillomotorischen Kerngebiet der Taube auf die Aktivität der Irismuskulatur. Pfluegers Arch 315:291 – 307
Pierau F-K, Torrey P, Carpenter DO (1974) Mammalian cold receptor afferents: role of an electrogenic sodium pump in sensory transduction. Brain Res 73:156 – 160
Pierau F-K, Torrey P, Carpenter DO (1975) Effect of ouabain and potassium-free solution on mammalian thermosensitive afferents in vitro. Pfluegers Arch 359:349 – 356
Pierau F-K, Klee MR, Klussmann FW (1976) Effect of temperature on postsynaptic potentials of cat spinal motoneurones. Brain Res 114:21 – 34
Pierau F-K, Wurster RD, Neya T, Yamasato T, Ulrich J (1980) Generation and processing of peripheral temperature signals in mammals. Int J Biometeorol 24:243 – 252
Poulos DA, Benjamin RM (1968) Response of thalamic neurons to thermal stimulation of the tongue. J Neurophysiol 31:28 – 43
Poulos DA, Lende RA (1970a) Response of trigeminal ganglion neurons to thermal stimulation of oral-facial regions. I. Steady-state response. J Neurophysiol 33:508 – 517
Poulos DA, Lende RA (1970b) Response of trigeminal ganglion neurons to thermal stimulation of oral-facial region. II. Temperature change response. J Neurophysiol 33:518 – 526
Poulos DA, Molt JT (1976) Response of central trigeminal neurons to cutaneous thermal stimulation. In: Zotterman Y (ed.) Sensory functions of the skin in primates. Pergamon, Oxford New York Toronto, pp 263 – 283
Price DD, Browe AC (1975) Spinal cord coding of graded non-noxious and noxious temperature increases. Exp Neurol 48:201 – 221
Price DD, Dubner R, Hu JW (1976) Trigeminothalamic neurons in nucleus caudalis re-

sponsive to tactile, thermal and nociceptive stimulation of monkey's face. J Neurophysiol 39:936–953
Proppe DW (1978) Effect of skin temperature on leg blood flow-core temperature relationship. Physiologist 21:94
Proppe DW, Brengelmann GL, Rowell LB (1976) Control of baboon limb flow and heart rate-role of skin vs. core temperature. Am J Physiol 231:1457–1465
Puschmann S, Jessen C (1978) Anterior and posterior hypothalamus: Effects of independent temperature displacements on heat production in conscious goats. Pfluegers Arch 373:59–68
Quazzani El T, Mei N (1979) Mise en évidence électrophysiologique des thermorécepteurs vagaux dans la région gastro-intestinale. Leur rôle dans la régulation de la motricité digestive. Exp Brain Res 34:419–434
Raths P, Hensel H (1967) Cutane Thermorezeptoren bei Winterschläfern. Pfluegers Arch 293:281–302
Rautenberg W (1971) The influence of the skin temperature on thermoregulatory system of pigeons. J Physiol (Paris) 63:396–398
Rautenberg W, Necker R, May ,B (1972) Thermoregulatory responses of the pigeon to changes of the brain and the spinal cord temperature. Pfluegers Arch 338:31–42
Rautenberg W, May B, Necker R, Rosner G (1978) Control of panting by thermosensitive spinal neurons in birds. In: Piiper J (ed) Respiratory function in bird, adult and embryonic. Springer, Berlin Heidelberg New York, pp 204–210
Rawson RD, Quick KP (1970) Evidence of deep-body thermoreceptor response to intra-abdominal heating of the ewe. J Appl Physiol 28:813–820
Rawson RO, Quick KP (1972) Localization of intra-abdominal thermoreceptors in the ewe. J Physiol (Lond) 222:665–677
Rawson RO, Stolwijk JAJ, Graichen H, Abrams R (1965) Continuous radiotelemetry of hypothalamic temperatures from unrestained animals. J Appl Physiol 20:321–325
Reaves TA, Heath JE (1975) Interval coding of temperature by CNS neurones in thermoregulation. Nature 257:688–690
Rexed B (1952) The cytoarchitectonic organization of the spinal cord in the cat. J Comp Neurol 96:415–495
Richards SA (1970) The role of hypothalamic temperature in the control of panting in the chicken exposed to heat. J Physiol (Lond) 211:341–358
Richards SA (1971) The significance of changes in the temperature of the skin and body core of the chicken in the regulation of heat loss. J Physiol (Lond) 216:1–10
Richards SA (1975) Thermal homeostasis in birds. Symp Zool Soc London 35:65–96
Richards S, Avery P (1978) Central nervous mechanisms regulating thermal panting. In: Piiper J (ed) Respiratory function in birds, adult and embryonic. Springer, Berlin Heidelberg New York, pp 196–203
Riedel W (1976) Warm receptors in the dorsal abdominal wall of the rabbit. Pfluegers Arch 361:205–206
Riedel W, Siaplauras S, Simon E (1973) Intra-abdominal thermosensitivity in the rabbit as compared with spinal thermosensitivity. Pfluegers Arch 340:59–70
Rosner G (1978) The influence of thermal stimulation of the spinal cord and skin on the activity of hypothalamic units. J Comp Physiol 126:151–156
Sakata Y (1979) Effects of pyrogen on the medullary temperature-responsive neurone of rabbits. Jpn J Physiol 29:585–596
Sand A (1938) The function of the ampullae of Lorenzini with observations on the effect of temperature on sensory rhythms. Proc Roy Soc Lond [Biol] 125:524–553
Schäfer K, Braun HA, Bade H, Hensel H (1979) EDTA-induced burst discharge in cold fibres of the cat's nose. Pfluegers Arch [Suppl] 379:R40
Schmidt I (1976) Paradoxical changes of respiratory rate elicited by altering rostral brain stem temperature in the pigeon. Pfluegers Arch 367:111–113

Scott NR, van Tienhoven A (1974) Thermoregulatory responses of poultry to local heating and cooling of the hypothalamus. Am Soc Agricult Eng 74:5512
Simon E (1972) Temperature signals from skin and spinal cord converging on spinothalamic neurons. Pfluegers Arch 337:323 – 332
Simon E (1974) Temperature regulation: The spinal cord as a site of extrahypothalamic thermoregulatory functions. Rev Physiol Biochem Pharmacol 71:2 – 76
Simon E, Iriki M (1971) Sensory transmission of spinal heat and cold sensitivity in ascending spinal neurons. Pfluegers Arch 328:103 – 120
Simon E, Simon-Oppermann C (1979) Metabolic thermoregulatory responses to CNS cooling and to general hypothermia in the conscious Pekin duck. Pfluegers Arch [Suppl] 382:R27
Simon E, Hammel HT, Oksche A (1977) Thermosensitivity of single units in the hypothalamus of the conscious Pekin duck. J Neurobiol 8:523 – 535
Simon E, Klussmann FW, Rautenberg W, Kosaka M (1966) Kältezittern bei narkotisierten spinalen Hunden. Pfluegers Arch 291:187 – 204
Simon E, Simon-Oppermann C, Hammel HT, Kaul R, Maggert J (1976) Effects of altering rostral brain stem temperature on temperature regulation in the Adelie penguin, Pygoscelis Adeliae. Pfluegers Arch 362:7 – 13
Simon-Oppermann C, Jessen C (1977) Antidiuretic responses to thermal stimulation of hypothalamus and spinal cord in the conscious goat. Pfluegers Arch 368:33 – 37
Simon-Oppermann C, Martin R (1979) Mammalian-like thermosensitivity in the lower brainstem of the Pekin duck. Pfluegers Arch 379:291 – 293
Simon-Oppermann C, Simon E, Jessen C, Hammel HT (1978) Hypothalamic thermosensitivity in conscious Pekin ducks. Am J Physiol 235:130 – 140
Simon-Oppermann C, Hammel HT, Simon E (1979) Hypothalamic temperature and osmoregulation in the Pekin duck. Pfluegers Arch 378:213 – 221
Snapp BD, Heller HC, Gospe SM Jr (1977) Hypothalamic thermosensitivity in California quail (Lophortyx californicus). J Comp Physiol 117:345 – 357
Speulda E, Wünnenberg W (1977) Thermosensitivity of preoptic neurones in a hibernator at high and low ambient temperatures. Pfluegers Arch 370:107 – 110
Spray DC (1974a) Characteristics, specificity, and efferent control of frog cutaneous cold receptors. J Physiol (Lond) 237:15 – 38
Spray DC (1974b) Metabolic dependence of frog cold receptor sensitivity. Brain Res 72:354 – 359
Spray DC (1975) Effect of reduced acclimation temperature on responses of frog cold receptors. Comp Biochem Physiol [A] 50:391 – 397
Steen I, Steen JB (1965) The importance of the legs in the thermoregulation of birds. Acta Physiol Scand 63:285 – 291
Stitt JT (1976) The regulation of respiratory evaporative heat loss in the rabbit. J Physiol (Lond) 258:157 – 172
Terashima S, Goris RC, Katsuki Y (1970) Structure of warm fiber terminals in the pit membrane of vipers. J Ultrastruct Res 31:494 – 506
Trevino DL (1976) The origin and projection of a spinal nociceptive and thermoreceptive pathway. In: Zotterman Y (ed) Sensory functions of the skin in primates. Pergamon, Oxford New York Toronto, pp 367 – 377
Trevino DL, Carstens E (1975) Confirmation of the location of spinothalamic neurons in the cat and monkey by the retrograde transport of horseradish peroxidase. Brain Res 98:177 – 182
von Düring M (1974) The radiant receptor and other tissue receptors in the scales of the upper jaw of Boa constrictor. Z Anat Entwicklungsgesch 145:299 – 319
Waites GMH (1962) The effect of heating the scrotum of the ram on respiration and body temperature. Q J Exp Physiol 47:314 – 323

Willis WD, Kenshalo DR Jr, Leonard RB (1979) The cells of origin of the primate spinothalamic tract. J Comp Neurol 188:543–574
Wünnenberg W, Brück K (1970) Studies on the ascending pathways from the thermosensitive region of the spinal cord. Pfluegers Arch 321:233–241
Wünnenberg W, Hardy JD (1972) Response of single units of the posterior hypothalamus to thermal stimulation. J Appl Physiol 33:547–552
Wünnenberg W, Merker G, Speulda E (1976) Thermosensitivity of preoptic neurones in a hibernator (golden hamster) and a nonhibernator (guinea pig). Pfluegers Arch 363:119–123
Wyss CR, Brengelmann GL, Johnson JM, Rowell LB, Niederberger M (1974) Control of skin blood flow, sweating, and heart rate: role of skin vs. core temperature. J Appl Physiol 36:726–733

A Review of the Auditory Physiology of the Reptiles

G. A. Manley

Institut für Zoologie, Technische Universität München, Lichtenbergstrasse 4, D-8046 Garching

1 Introduction

The study of the auditory physiology of reptiles has a relatively long history, but only began in a systematic way in 1956 with the publication of the earliest of Wever's investigations of the cochlear microphonic in reptiles. The long series of experiments which were subsequently undertaken by Wever and his colleagues have been recently conveniently brought together with the publication of Wever's book *The Reptile Ear* (1978). In the last 10 years, neurophysiological studies at various levels of the auditory system (primarily, however, lower levels), have appeared and produced in a relatively short time a good basis for the discussion of mechanisms. Certainly, a great difference can be noted today between our increasing unterstanding in this field and the paucity of data which existed in 1960 when McGill could say

Disagreement exists ... as to whether the hearing organs of certain modern reptiles are vestigial or rudimentary. The present state of knowledge of hearing in ... reptiles is not commensurate with the importance of these classes in the study of the evolution of the sense of hearing (McGill 1960).

Two main themes dominate the motivation underlying present research in this field. The first, and historically older, theme is a fundamental interest in the evolution and systematics of the reptile ear. To what extent can the diverse ear structures of modern reptiles provide clues, firstly, as to the origins of the ear of terrestrial vertebrates and, secondly, as to the relationships of these vertebrate groups to one another? The second theme is that of using the reptile ear and central pathways as a "simple" model in order to indirectly learn more about the hearing systems of mammals and man. This comparative approach is based on the well-founded assumption that because these hearing systems have a common ancestry in the lateral-line system of fish and the actual ear of early stem reptiles, they will share many functional principles. The relative ease of investigation and structural diversity of the reptile inner ear provide, as I hope to show below, fruitful ground for approaching these important questions.

Miller (1980) lists the four living orders of reptiles in order of the apparent state of development of the cochlear duct and basilar papilla. The most primitive cochlear duct is listed first:

Order Rhynchocephalia (the tuatara, *Sphenodon*)
Order Testudines (turtles, tortoises, terrapins)
Order Squamata
    Suborder Serpentes (snakes)
    Suborder Amphisbaenia (amphisbaenids)
    Suborder Lacertilia (lizards)
Order Crocodylia (crocodiles, alligators, gavials)

All of these orders are traceable to Triassic times (Carroll 1969). The suspected earlier anapsid ancestry of the turtles is not yet established. The central stock of Permocarboniferous reptiles, the Captorhinomorphs, gave rise to the thecodont ancestors of the Crocodilia and the eosuchian ancestors of the Squamata and Rhynchocephalia. Thus the four orders have a history of at least 200 million years of separate evolution.

2 Middle Ear Structure and Physiology

The external ear of reptiles is often hardly present and, except in special cases (see Wever 1978), plays no significant role in the filtering or funneling of the sound signal at the periphery. The middle ear, in contrast, differs in some important respects from that of mammals, strongly influences the characteristics of the ear's frequency response and needs to be discussed.

Although there are differences between different orders and families of reptiles in the detail of the middle ear structures (Wever 1978), the net functional result is rather similar in most cases. Basically, the middle ear consists of a relatively thin, often almost transparent eardrum (sometimes thick or absent) which is closely attached to the various processes of a flexible extracolumella (Fig. 1). This extracolumella directly joins a columella, which at its inner end widens out into a footplate more or less covering the oval window of the inner ear.

As in all terrestrial vertebrate ears, the middle ear functions as an impedance-matching device between the air and the fluids of the inner ear. This impedance match is on average as successful as in mammals. First of all, the match is partly due to the area ratio between the eardrum and columella footplate. This essentially concentrates much of the pressure on the eardrum on to the much smaller footplate. In the Tokay gecko, this ratio is between 1:45 and 1:60 (Manley 1972a; Wever 1978). In the iguanid lizard *Crotaphytus collaris* it is about 1:20

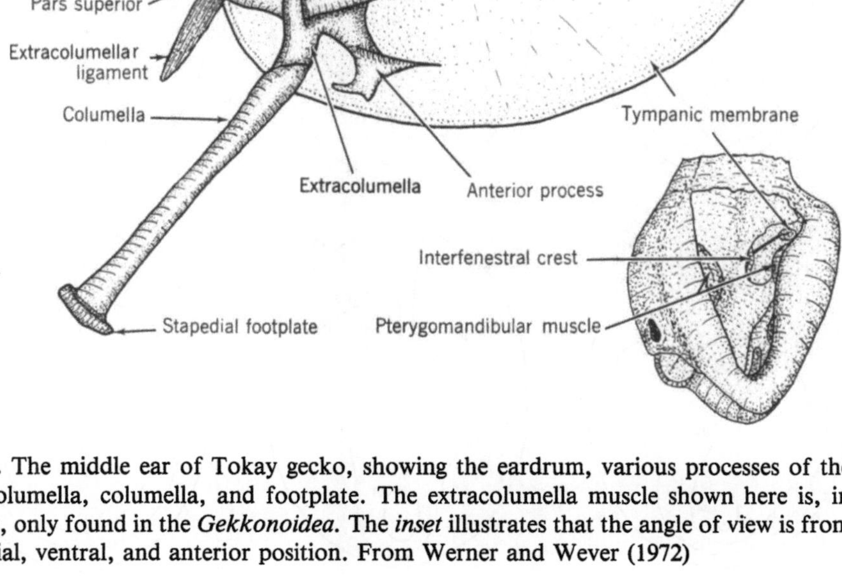

Fig. 1. The middle ear of Tokay gecko, showing the eardrum, various processes of the extracolumella, columella, and footplate. The extracolumella muscle shown here is, in lizards, only found in the *Gekkonoidea*. The *inset* illustrates that the angle of view is from a medial, ventral, and anterior position. From Werner and Wever (1972)

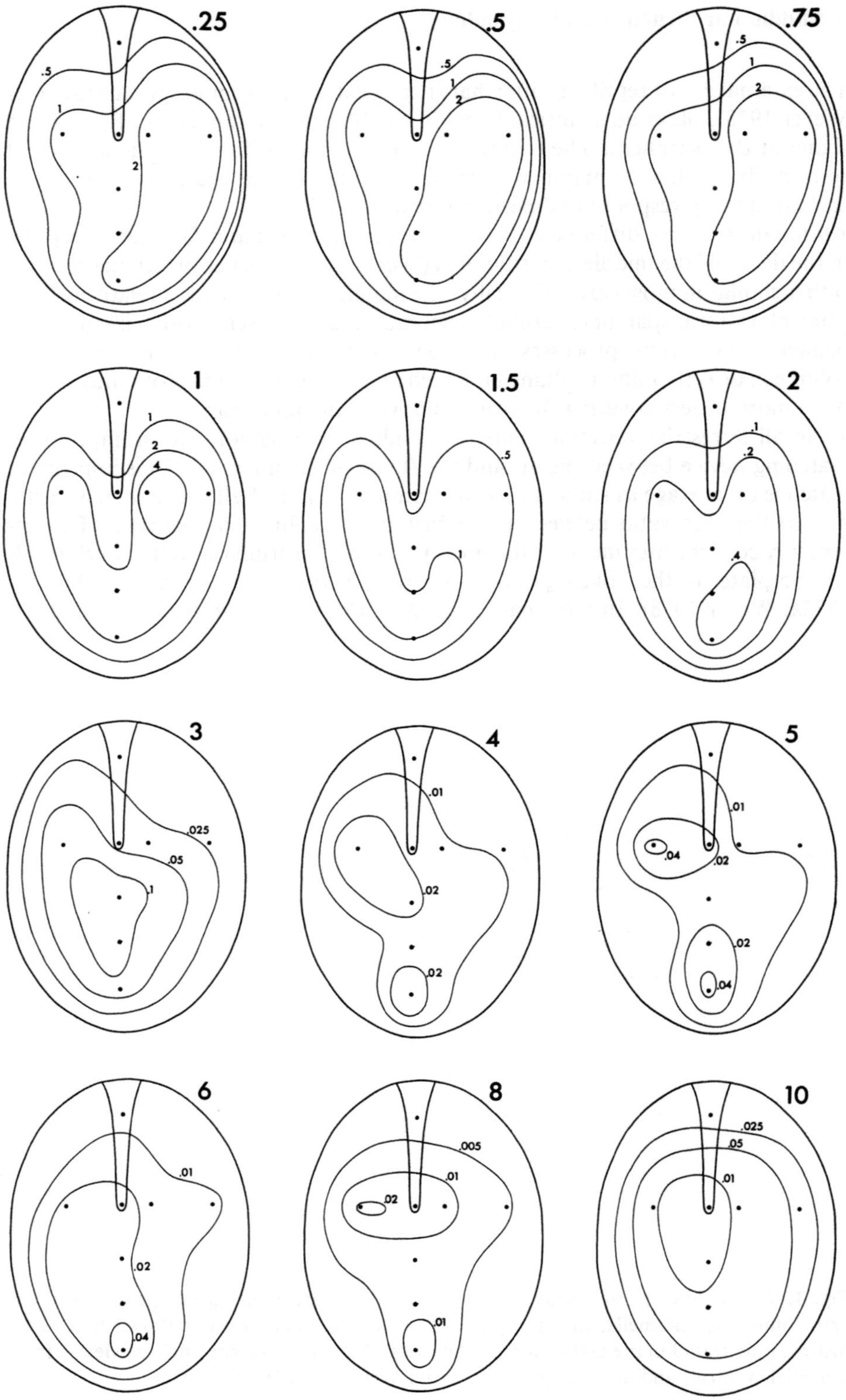
.25
.5
1
2
.5
.5
1
2
.75
.5
1
2
1
1
2
4
1.5
.5
1
2
.1
.2
.4
3
.025
.05
.1
4
.01
.02
.02
5
.01
.04
.02
.02
.04
6
.01
.02
.04
8
.005
.01
.02
.01
10
.025
.05
.01

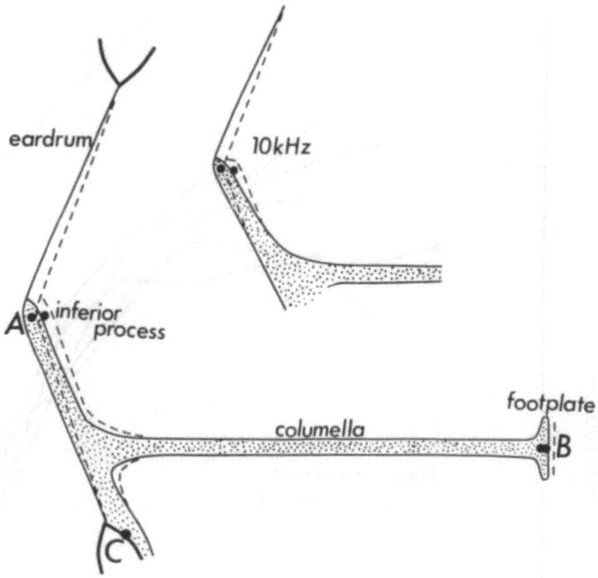

Fig. 3. Diagrammatic section through the middle ear of a gecko illustrating the operation of the middle ear lever. The inferior process (pars inferior in Fig. 1) of the extracolumella (*A* to *C*) hinges at point *C* at low frequencies, and the displacement of the columella and footplate (*B*) is proportionately reduced from that at point *A*. At high frequencies (10 kHz, *inset*) the inferior process flexes, increasing the difference in displacement amplitude between point **A** and the footplate. From Manley (1973)

(Wever 1978). The effective area of the drum involved, however, varies with the changing vibration pattern at different frequencies (Fig. 2; Manley 1972a, c). In addition, the typical system for connecting the extracolumella and columella (Fig. 3) provides a second-order lever in the sound-transmission chain, again increasing the force available at the inner ear. A lever action has been demonstrated in *Crotaphytus* by Wever and Werner (1970) using cochlear microphonic measurements and in geckos by Manley (1972a, b) using Mössbauer measurements of the velocity of various components of the middle ear. Thus although the nonmammalian middle ear pattern has an independent origin and a different structure from that of mammals, the components contributing to impedance

Fig. 2. Isoamplitude contours of the eardrum of the Tokay gecko at 12 frequencies (0.25 – 10 kHz), measured by the Mössbauer technique. *Small numbers*, displacement amplitude of each contour in microns. *Dots*, measurement locations. Each contour joins points having the same displacement amplitude at 100 dB SPL. The coupling of eardrum displacement to the columella system is frequency dependent. From Manley (1972a)

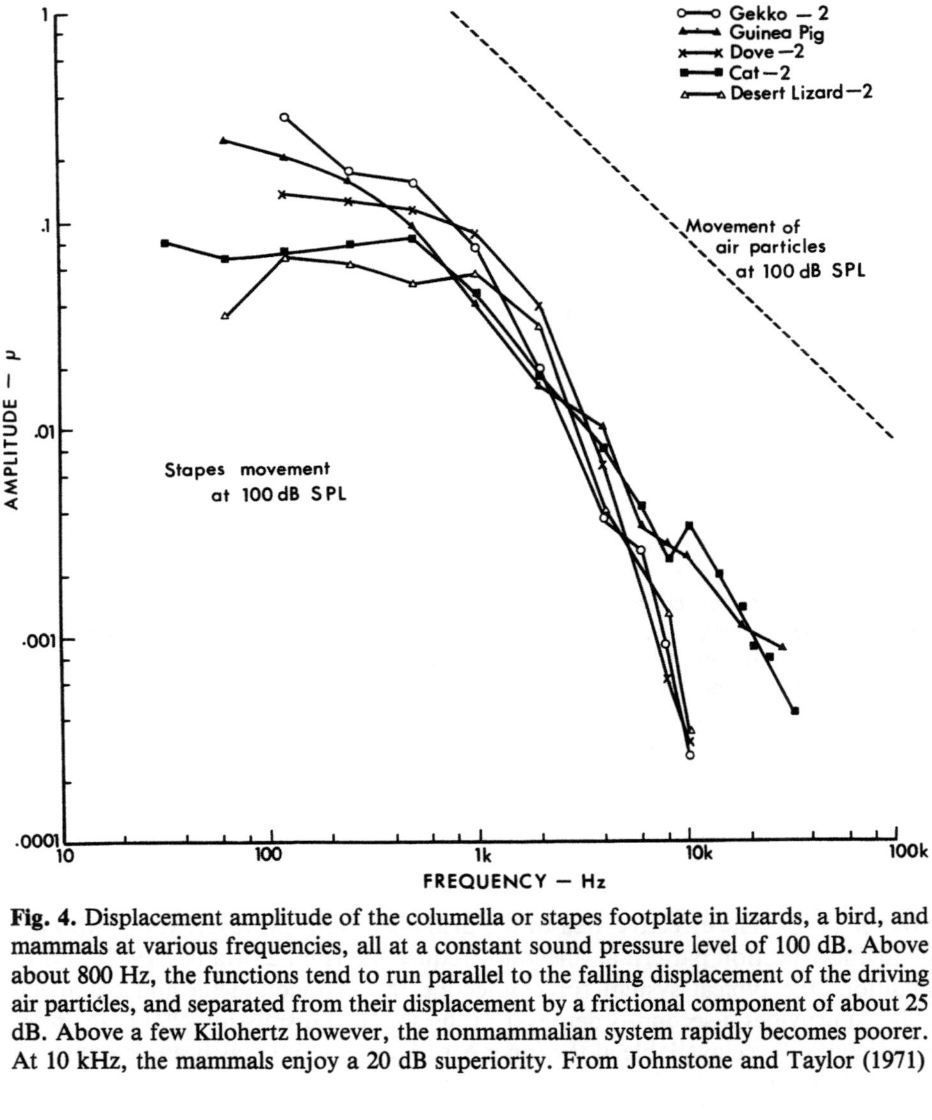

Fig. 4. Displacement amplitude of the columella or stapes footplate in lizards, a bird, and mammals at various frequencies, all at a constant sound pressure level of 100 dB. Above about 800 Hz, the functions tend to run parallel to the falling displacement of the driving air particles, and separated from their displacement by a frictional component of about 25 dB. Above a few Kilohertz however, the nonmammalian system rapidly becomes poorer. At 10 kHz, the mammals enjoy a 20 dB superiority. From Johnstone and Taylor (1971)

matching (area and lever ratio) are the same and, indeed, at some frequencies equally effective (Fig. 4). In *Gekko gecko*, the largest transmission loss on middle ear removal and direct stimulation of the oval window was found to be 35 dB in the midfrequency range; in *Eublepharis*, the loss was 53 dB (Werner and Wever 1972), a range of values very similar to those available for mammals. At low frequencies (Fig. 4) the resistive or frictional loss in displacement as compared to that of the air particles is approximately the same as that of mammals. There is, however, a substantial difference in the frequency response characteristics of the two kinds of middle ears. The mammalian middle ear has a broader frequency pass characteristic (Johnstone and Taylor 1971; Manley et al. 1972; Manley and Johnstone 1974) than that of nonmammals. This difference is most pronounced above 4 kHz. Moffat and Capranica (1978) found that the middle-ear transfer function of the turtle *Pseudemys* fell off rapidly at frequencies above 650 Hz. While some of the loss of transmission at frequencies higher than the optimum (which lies within an octave of 1 kHz) may be due to a change in the input impedance of the inner ear, one component of the loss is certainly due to the way in which the middle-ear lever is constructed. The extracolumella is a flexible structure, presumably at least in part to allow it to be less liable to damage by contact with external objects from which it receives little if any protection from an external ear. The flexibility also allows the junction with the extracolumella – columella shaft connecting to the footplate to flex and enable pistonlike motion of the footplate (Fig. 3). Perhaps as a result of this flexibility the inferior process of the extracolumella, which is the main transmitter of energy from the eardrum to the columella, flexes at high frequencies (>4 kHz), thus reducing the force available to the inner ear. It seems that this inherent requirement for flexibility has thus placed a profound limitation on the frequency range of the nonmammalian ear (Fig. 5; Manley 1973). Of course, this limits the range of frequencies available to these species for use in auditory communication. Even the complex acoustic signaling in birds is limited to much lower frequencies than the average mammal can hear (Manley 1973). It is difficult to estimate to what extent this factor has exerted an influence on the modest development of such communication behavior in reptiles.

In some cases, modifications of the feeding mechanism have led to profound changes in middle ear structures, reducing auditory sensitivity (Berman and Regal 1967; Wever 1978). Details of the structural variations can be found in Olson (1966) and in Wever (1978).

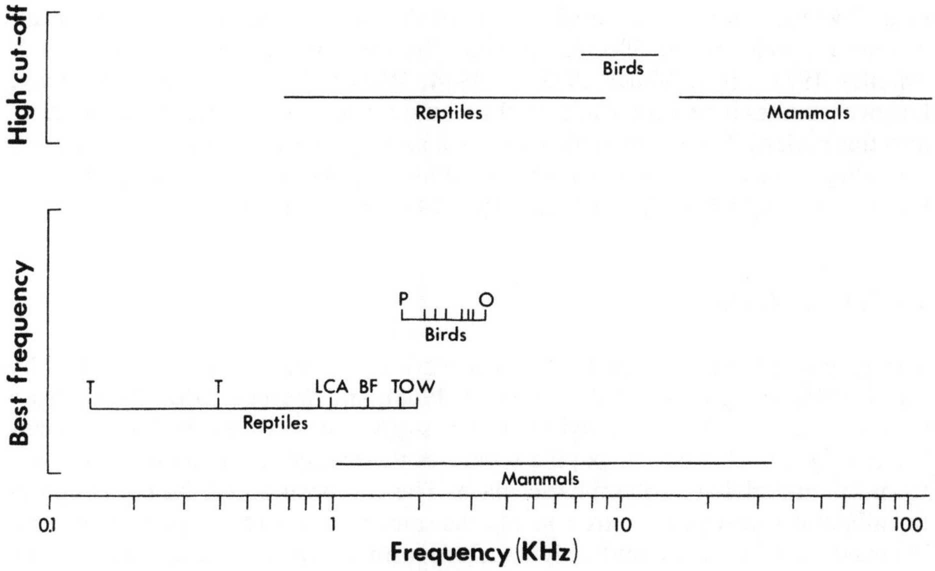

Fig. 5. Distribution of best frequency (or most-sensitive frequency) and high-frequency sensitivity (high cut-off is the highest "perceptible" frequency at a sound pressure level of 75 dB) in reptiles, birds and mammals. *P*, pigeon; *O*, barn owl; *T*, Box turtle; *L*, leopard lizard; *CA*, caiman; *BF*, blue fence lizard; *TO*, Tokay gecko; *W*, Western banded gecko. From Manley (1971)

A further factor in middle-ear physiology which has recently been under discussion in the case of birds (and under similar conditions in the insects) is the open channel between the middle-ear cavities (Coles et al. 1980; Hill and Boyan 1976, 1977; Hill et al. 1980), which allows the middle ear to operate as a pressure-gradient receiver. It has been demonstrated in the Japanese quail that sound energy transmitted through the tympanum can strongly interfere with the response of the contralateral middle ear, depending on the phase and intensity relationships involved. These latter factors are in turn dependent on the frequency and directionality of the sound source. These findings are interesting in that they provide a peripheral basis for the encoding of sound directionality information at frequencies where the head itself provides very little sound shadow. The question of how small nonmammals, which only have low-frequency hearing, could use the directionality information available "conventionally" (i.e., sound intensity shadow, phase and time differences) has always been a difficult one. It will be necessary to carry out measurements in reptile ears before the significance of this interaural interaction can be accurately assessed.

3 Inner Ear Structure and Physiology, Including the Auditory Nerve

One of the most remarkable features of the reptilian ear, seen most profoundly in the lizards, is the systematic structural diversity of the inner ear. As has been previously noted, this diversity provides an unparalleled series of "natural experiments" which, with careful consideration of the questions to be asked and the appropriate species to be selected, provide the experimenter with ideal material (Manley 1973, 1977; Miller 1973a,b, 1974; Wever 1978). This very diversity, however, makes it necessary to introduce an overview of the anatomical features into this review. The ear of snakes will be largely ignored, as single-neuron work providing a basis for discussing their auditory physiology is lacking (but see Hartline 1971a,b; Hartline and Campbell 1969; Miller 1978b; Wever 1978).

3.1 General Anatomy

The general anatomy of the inner ear in various reptiles has been described by Baird (1960, 1970, 1974), Miller (1966a,b, 1968), and Wever (1978). The auditory cells, or "hair" cells are located with their supporting cells on the basilar membrane (Fig. 6). This basilar papilla separates the overlying endolymphatic space from the underlying perilymphatic space. The composition of the perilymph in the alligator lizard and cat are roughly the same but, due to sampling problems, Peterson et al. (1978) could only conclude with regard to endolymph that the potassium concentration is high and the sodium concentration is low, as in mammals (Johnstone et al. 1963; Johnstone and Sellick 1972).
The upper surfaces of the sensory cells are also exposed to a small positive potential (of mean value +16 mV in the alligator lizard, Weiss et al. 1978a) which would tend to increase the force driving potassium ions through the basilar

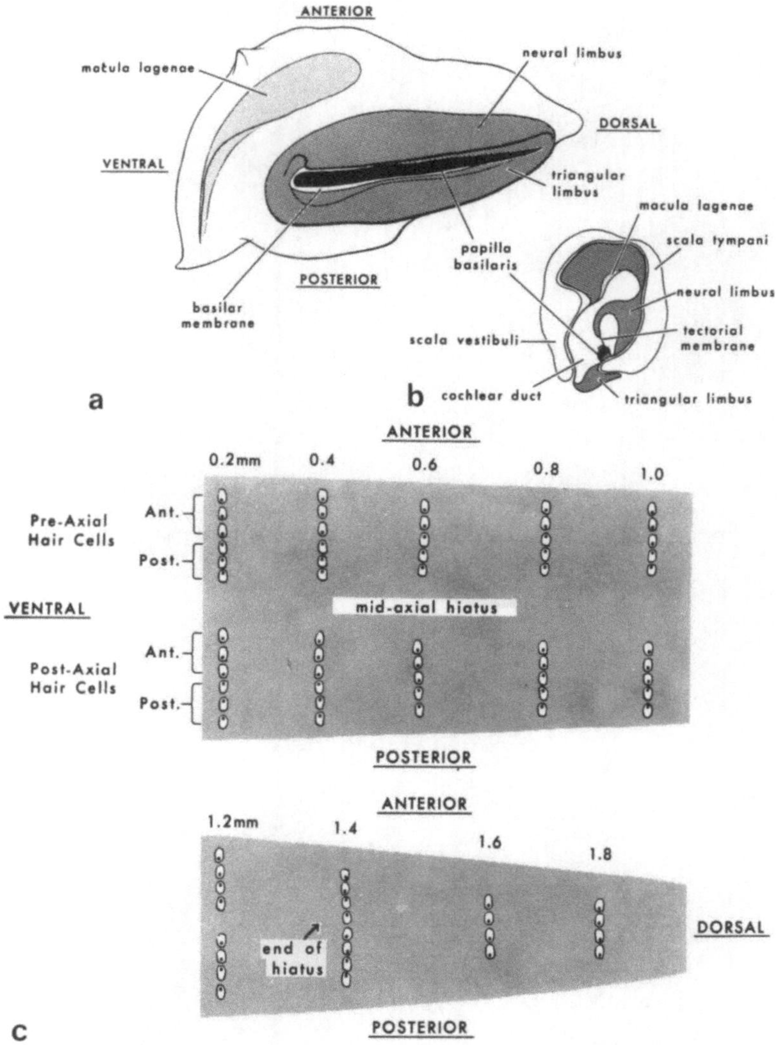

Fig. 6. a Part of the left cochlear duct of *Gekko gecko* seen from the lateral aspect. The supporting limbus surrounds the basilar papilla, and on the neural side arches up to support the tectorial membrane. **b** Cross-sectional view of the cochlear duct in its ventral portion. The periotic sac surrounds the duct and is divided into the scala tympani and vestibuli. **c** Diagrammatic representation of the regular orientation of the hair cell kinocilia in *Gekko gecko* in different areas of the papilla. The *dot* at one end of each cell represents the kinocilium. The extreme dorsal, narrow end is unidirectional (cf. Fig. 20). From Miller (1973a)

papilla into the perilymph or blood stream. These factors are similar to, but smaller than those described for mammals, and may play a role in auditory transduction as described by Davis (1968).

The sensory cell itself is structurally similar to all hair cells of the acousticolateralis system (Fig. 7; see, e.g., Angelborg and Engström 1973; Bagger-Sjöbäck

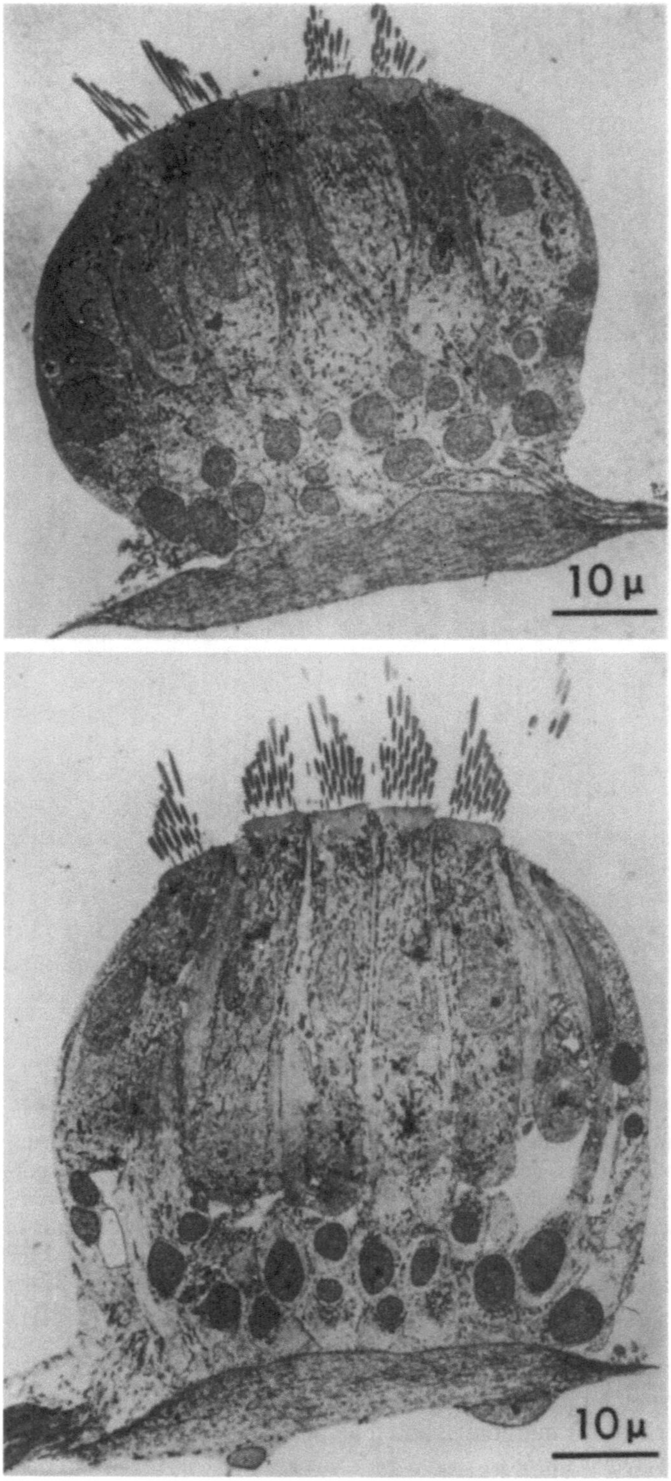

Fig. 7. Cross section through the apical part of the basilar papilla of *Calotes versicolor* showing short-ciliated hair cells of a unidirectionally oriented area and thickened basilar membrane (*above*). *Below*: long-ciliated hair cells of the basal area of the same papilla, in a bidirectionally oriented area. From Bagger-Sjöbäck (1976)

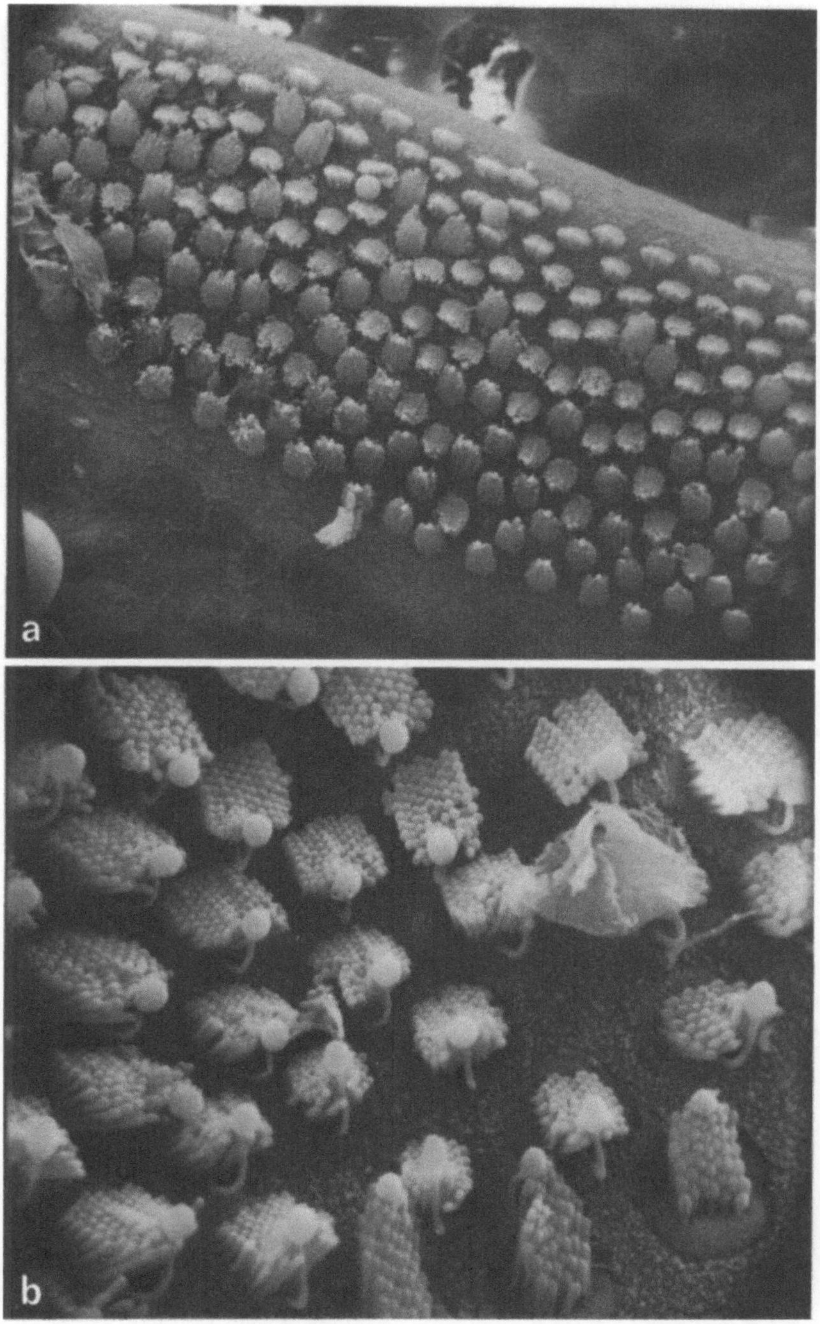

Fig. 8. **a** Scanning electron micrograph (*SEM*) of hair cell bundles (seen almost from above) of the dorsal region of the papilla of *Tupinambis teguixin* (Teiidae) illustrating the appearance of a bidirectionally oriented group of hair cells. A pattern of orientation here is difficult to determine, although groups of cells oriented in one direction tend to alternate with groups oriented in the other direction. **b** Higher-power SEM of a group of hair cells of the papilla of *Ameiva ameiva* (Teiidae), which are mostly oriented posteriorly. The three lower cells have the oppositive orientation. The elongated kinocilium and kinocilial bulb attached to the longest stereocilia can be clearly see. From Miller (1973b)
a × 1180; **b** × 4800

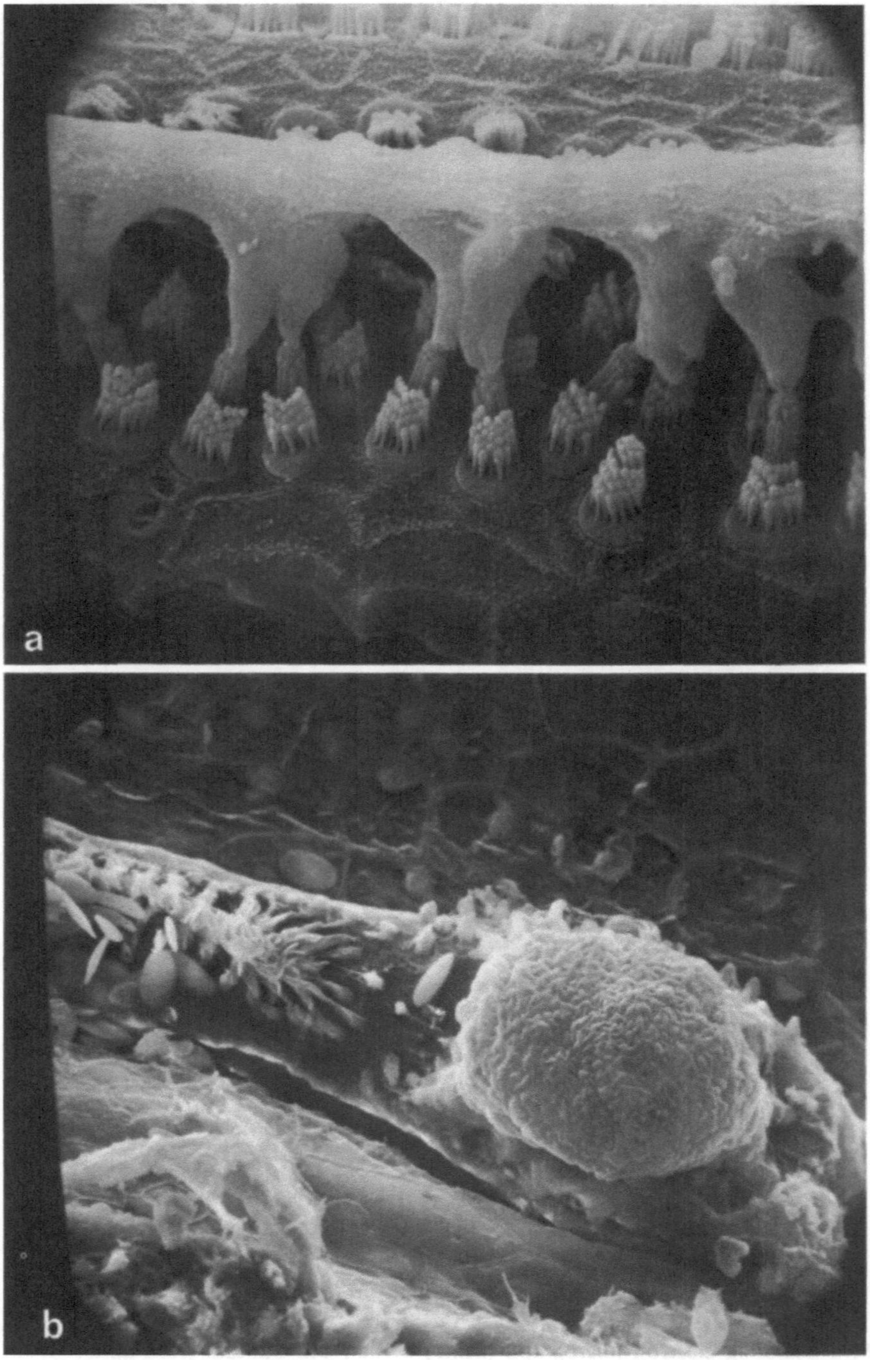

Fig. 9 a, b. Tectorial membrane structures unique to lizards. **a** Sallets of *Eublepharis* (Eublepharidae). The sallets are connected by a thick band of tectorial tissue. Miller (1973b). **b** Large mass of tectorial material covering the hair cells of the apical expansion of the basilar papilla of *Eumeces gilberti* (Scincidae). Tectorial material in the form of sallets covers the nonapical hair cells. From Miller (1974) **a** × 2300; **b** × 610

1976; Engström 1967; Smith 1968; Wersäll and Flock 1967). There is a variably large number of stereocilia on the free surface, the longest of which are usually connected to the kinocilium. These hair cells are polarized in their directional sensitivity by the location of the kinocilium, being preferentially sensitive to cilial displacement in one direction (Flock et al. 1973). The orientation patterns of the cilia in different papillae vary, being unidirectional and abneural in snakes, turtles, and crocodilians and showing both unidirectional and bidirectional areas in the lizards (Fig. 8). The cilia of some cells are connected to a tectorial membrane, which itself may take on various fantastic forms, as described by Miller (1973a,b, 1974) and Wever (1978) (Fig. 9). The length of the stereocilia is affected by the presence or absence of the tectorial membrane in that particular region (see, e.g., Bagger-Sjöbäck and Wersäll 1973). Cilia not connected to a tectorial membrane tend to be longer and vary in length, in hair cells in different locations, along the papilla. In all reptiles, it appears also as if there is a network of tectorial-like material overlying the supporting cells and connected to the tectorial membrane proper (Miller 1978b).

Differences also exist in relation to the presence or absence of efferent innervation. It seems that all auditory papillae, except the basilar papilla of anuran amphibians, the bidirectional parts of all lizard papillae, and possibly the unidirectional regions of some lizard species (Bagger-Sjöbäck 1976) receive efferent nerve terminals (Miller 1978b). Details of the actual innervation patterns of both afferent and efferent systems are generally lacking. In *Calotes,* Bagger-Sjöbäck (1976) found that the nerve fibers (all afferent) are undivided, each contacting only one hair cell. Each hair cell received, on average, three to four terminals.

3.2 Turtles

The turtle ear is thought to represent many features characteristic of *ancestral* reptile ears. These are a low density of hair cells, unidirectional hair cell orientation, synapses between hair cells and both afferent and efferent nerve fibers, only one basic type of cytologically relatively unspecialized hair cell, and lastly, hair cells with a large number of stereocilia (90+) (Miller 1978b). In the turtle, a large number of hair cells are found in part of the basilar papilla which lies outside the basilar membrane on the limbic structure, in the form of a forked or T-shaped end or as a hooklike process (Fig. 10; Miller 1978b). The hair cells on the basilar membrane are connected to a tectorial membrane, whereas the hair cells on the limbus are connected with a gelatinous material and are presumably stimulated in a manner similar to the hair cells of the amphibian papilla of the amphibia.

The turtle basilar papilla also shows length variation for the cilia, the stereocilia being peripherally longer than those on cells in the center of the papilla (or the cells at the extremes are smaller; Miller 1978b). This may be related to their mode of stimulation.

The turtle papilla is generally large and contains many almost exclusively abneurally oriented hair cells (1100 – 1400 in the three species studied by Miller 1978b). The large dimensions and simple form of the elliptical basilar membrane lead to a relatively low frequency response. This fact was clear in the early cochlear microphonic data of Wever (1978) and was confirmed by the only

Fig. 10. Low-magnification (× 94) view from above of the basilar papilla of the turtle, *Pseudemys scripta*. Note the wide basilar membrane (*bm*), hair cells on the apical (*a*) and basal (*b*) limbus and artifactual break at *b*. The tectorial membrane is mostly missing in this preparation. *Ab l*, abneural limbus. From Miller (1978b)

behavioral audiogram obtained from a reptile, in the turtle *Chrysemys (Pseudemys) scripta* (Patterson 1966; see Fig. 11).

Recent electrophysiological recordings from auditory nerve fibers in a turtle by Paton et al. (1976) substantiate this finding. They studied more than 100 primary auditory units units in the red-eared turtle, *Pseudemys scripta,* and found that these units were most sensitive to frequencies of 80 – 450 Hz. These most sensitive frequencies for different units are called their characteristic frequencies or *CFs.* The sharpness of the tuning function for different units increased monotonically with CF. This sharpness of tuning is usually quantified by measuring the width of the bandwidth of frequencies to which the cell responds at a sound pressure level 10 dB more intense than the threshold at the CF. This $Q_{10\,dB}$ (or simply Q value) is obtained by dividing the CF by the bandwidth. Sharply tuned units thus have high Q values.

In response to clicks, Paton et al. found that turtle primary afferents often gave several action potentials, the intervals between which often corresponding to the reciprocal of the CF. This suggests that, as is usual, the basilar membrane swings through several oscillations after a click. In contrast to mammal units, however, the click-response latencies were shortest for condensation clicks, rather than for rarefaction clicks.

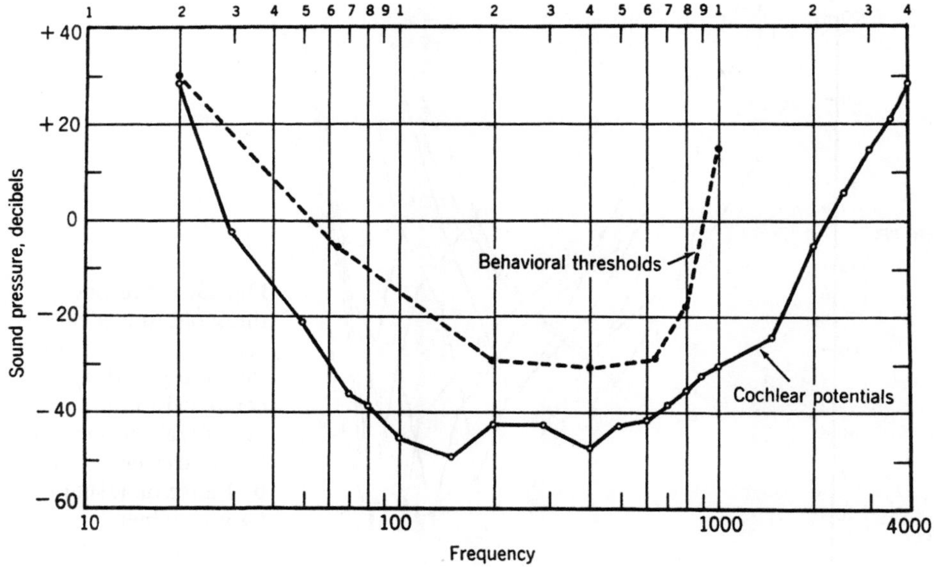

Fig. 11. Behavioral threshold and cochlear microphonic potential function (level, 0.1 μV) for the turtle *Chrysemys* (*Pseudemys*) *scripta*. The sound-pressure axis is in decibels with reference to 1 dyn/cm² or 0.1 N/m². From Wever (1978)

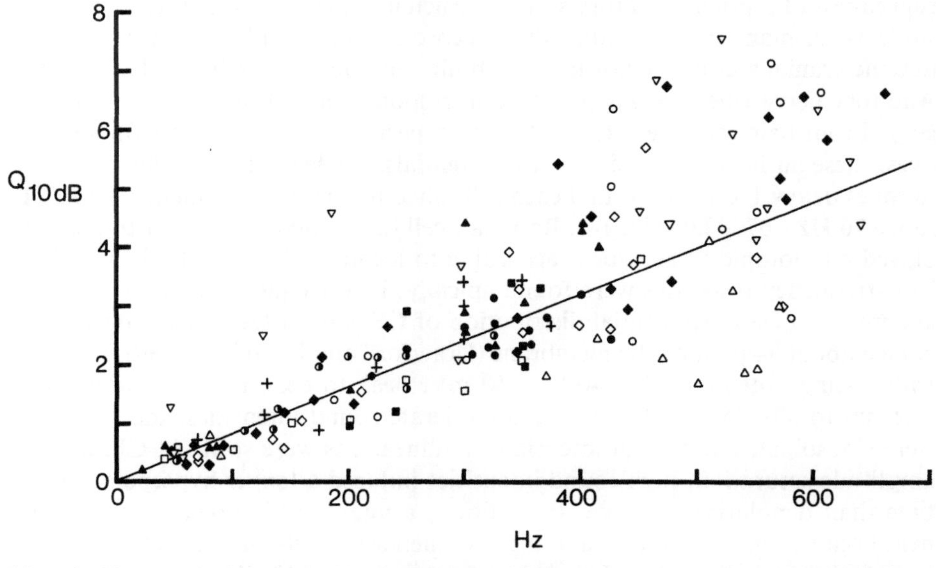

Fig. 12. Sharpness of tuning ($Q_{10\ dB}$) of single auditory nerve fibers in the turtle plotted against the CF of the fibers. Different symbols for 11 different animals. The frequency axis is linear. The line was drawn through the points by eye and represents a constant 10 dB bandwidth of 125 Hz for the tuning curves. From Crawford and Fettiplace (1980)

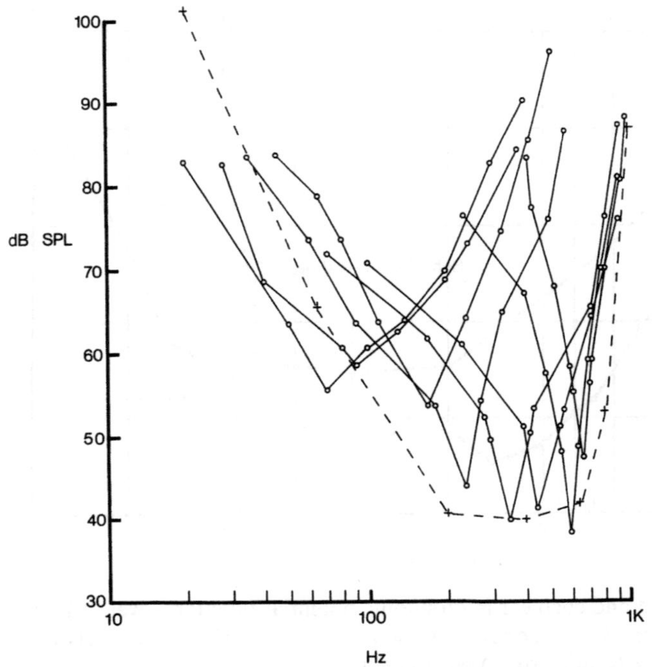

Fig. 13. Frequency-threshold tuning curves for six nerve fibers of the turle *Pseudemys* compared to the behavioral threshold for this species determined by Patterson (1966) (*crosses* joined by *dashed line*). From Crawford and Fettiplace (1980)

Working with primary fibers in the same species of turtle, Fettiplace and Crawford (1978) recorded from fibers with Q values of 0.2 – 7.5 (Fig. 12) with a CF range of 30 – 700 Hz (Fig. 13). Crawford and Fettiplace (1978, 1979, 1980) and Fettiplace and Crawford (1978) report intracellular recordings from basilar papilla hair cells in the turtle *Pseudemys scripta*. Illustrating the physiological ruggedness of reptilian auditory systems which makes them suitable experimental subjects for many investigations, these recordings were made after decapitation, midline cranial section, removal of the brain, and opening of the scala tympani! Auditory nerve fibers in this preparation respond for at least 4 h after the surgery. From hair cells identified after the experiment from injected fluorescent dyes, these authors recorded receptor potentials, graded with stimulus intensity, to tones below 1 kHz and found each cell tuned to a narrow frequency band (CF range 70 Hz – 670 Hz) (Fig. 14). Both hair cell tuning and nerve-fiber tuning displayed a tonotopic arrangement according to location along the basilar papilla. Low-frequency responses were found apically, high frequencies basally. On the assumption of an exponential distribution of CF with distance, each octave occupied about 94 μ along the membrane (Crawford and Fettiplace 1980). For cells with resting potentials of – 40 to – 50 mV, peak-to-peak response amplitudes were up to 30 – 45 mV. For low and moderate sound intensities, the responses were sinusoidal, but at high intensities nonlinearities were seen. At CF, despite amplitude saturation, high sound intensities induced a longer hyperpolarization time than depolarization time. In addition, a maintained depolarizing component became more pronounced at high frequencies which may reflect a rectification in the periodic signal. Loud, low-frequency tones built up approximately exponentially during the tone and elicited, in addition to the sinusoidal response at the stimulus frequency, an additional terminal "ringing" response (Fig. 15b), whose frequency varied with the CF of the cell, but which for a given cell was in-

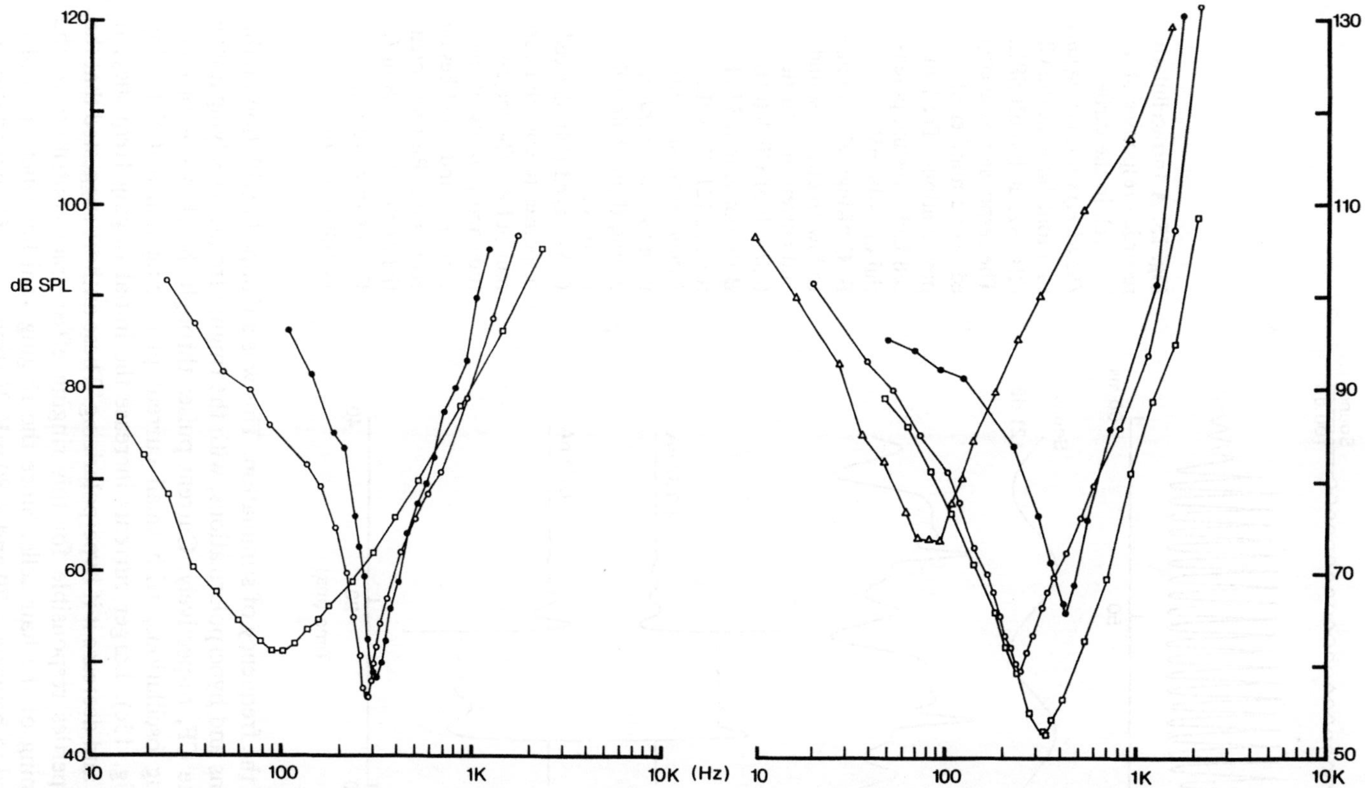

Fig. 14. Isoamplitude tuning curves for intracellular responses from seven turtle hair cells. The *ordinate* is the sound pressure (expressed in decibels re 20 μPa) required to produce a constant amplitude of 1 mV peak-to-peak response as a function of the stimulation frequency on the *abscissa*. Note a 10 dB difference between the left and right sound pressure axes. From Crawford and Fettiplace (1980)

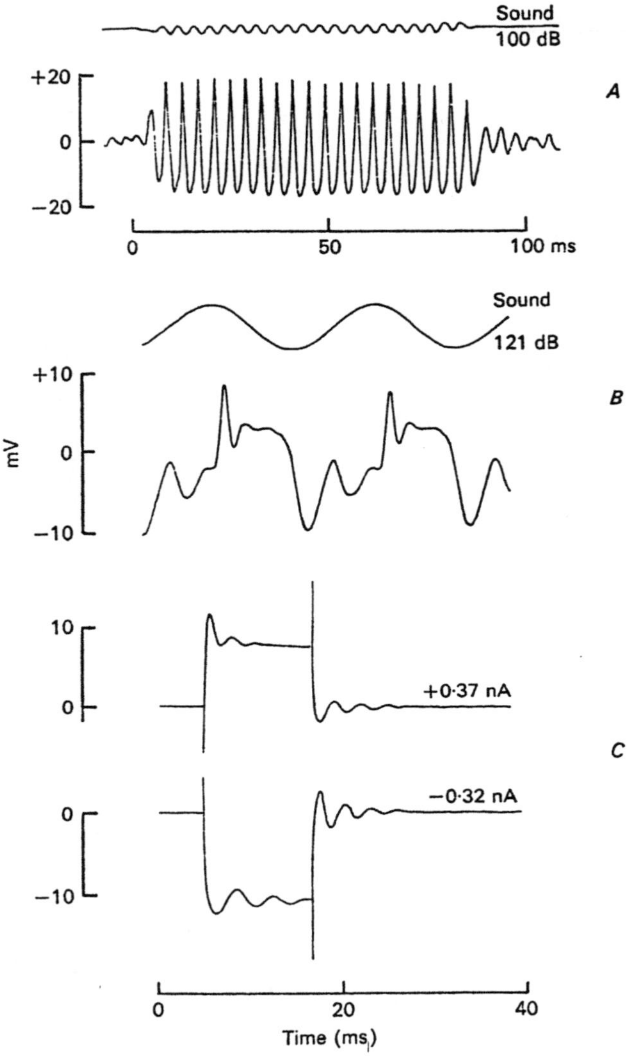

Fig. 15. A Intracellularly recorded responses of a hair cell in the turtle *Pseudemys scripta elegans* to a tone burst at the CF (250 Hz) at 100 dB SPL. The response was averaged over a number of presentations. The hair cell had a resting potential of −48 mV. **B, C** "Ringing" responses to low frequency sound and current in a turtle hair cell of CF 340 Hz. **B** Response to a 57 Hz tone at 121 dB SPL. *Upper trace,* stimulus monitor; *lower trace,* averaged (64) hair cell response. **C** Averaged (64) hair cell responses to constant current pulses. The magnitude and polarity of the current is indicated beside each trace. Resting potential of the cell, −48 mV. From Crawford and Fettiplace (1978)

dependent of the frequency of stimulation. This was a damped oscillation on the depolarizations and hyperpolarizations, with the ringing frequencies lying above and below the CF, respectively. Current pulses through the electrode produce similar ringing oscillations, with small current producing a near-CF ringing frequency (Fig. 15c). Larger currents increase the initial ringing frequency, if depolarizing, and decrease it if hyperpolarizing. These authors suggested that the hair cell properties responsible for this ringing effect may contribute to the frequency tuning of the hair cells. Since the ringing vanishes when the cell is hyperpolarized to between −70 and −80 mV, Fettiplace and Crawford suppose that a voltage-sensitive potassium conductance may be involved.

Using combined current injection and tone bursts, Crawford and Fettiplace (1979) estimated the hair cell reversal potential to be −3.5 ± 2.5 mV for cells with resting potentials of −30 to −40 mV. Thus they suggest that the transducer

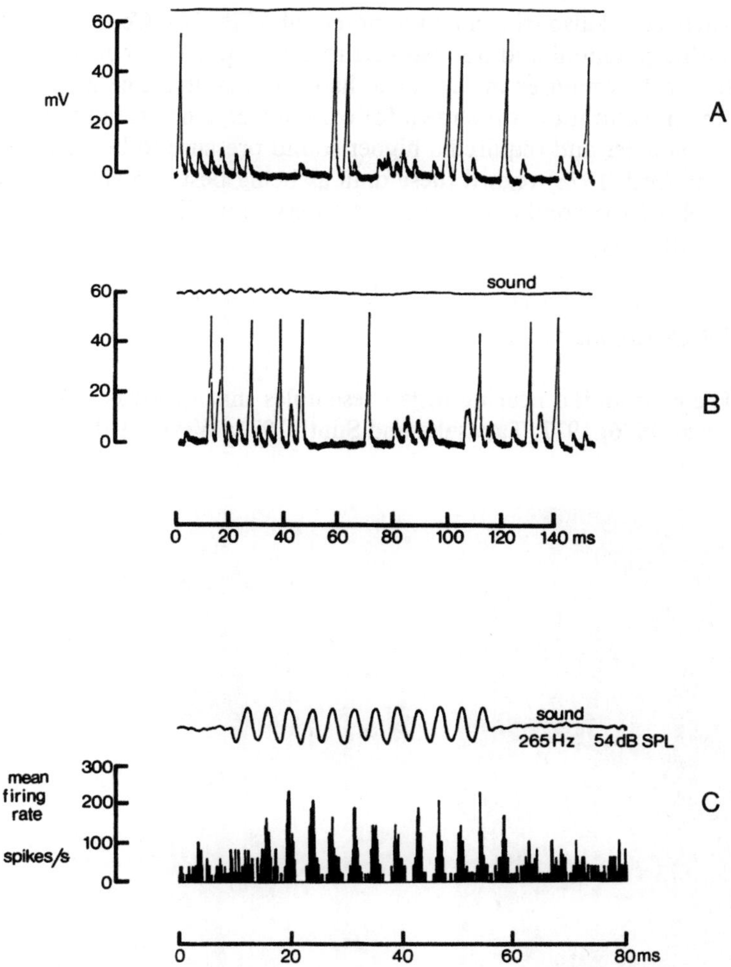

Fig. 16 A – C. Intracellularly recorded responses of a cell in the turtle auditory papilla presumed to be a terminal of an auditory nerve fiber. **A** Record in the absence of deliberate sound monitor. Voltage expressed relative to the cell's resting potential, in this case – 63 mV. Spontaneous potentials occur which trigger action potentials on seven occasions in this illustration. **B** As in **A**, but with sound stimulation in the early part. The pure tone was at CF (265 Hz) and 54 dB SPL. **C** PST histogram, to show phase-locking of impulses, made from 290 responses similar to that in **B**. From Crawford and Fettiplace (1980)

current flow could arise either by potassium ions moving across the endolymph face of the hair cell or by a nonselective cation conductance change in the hair cell membrane facing the perilymph.

In addition to recording from hair cells, these authors report recordings which probably originated in auditory nerve terminals. Here, resting potentials were near – 60 mV and the responses were a rectified form of the sound wave which could trigger action potentials. In addition, these units were characterized by the presence of randomly occurring brief depolarizations in the absence of sound which could also trigger action potentials (Fig. 16). Cells found which had a large resting potential and no response to sound were probably supporting cells. Hair cell responses in the turtle had a dynamic range at CF of at least 50 dB. Similar saturation was shown for off-CF frequencies, but the saturation response was smaller and required a higher sound pressure to be obtained. Fettiplace and Crawford (1978) regard these data as being best explained if the response is the result of the combination of two filters, both linear, separated by a saturating nonlinearity.

3.3 Crocodilia

The crocodilian hearing organ resembles that of birds (von Düring et al. 1974; Leake 1976, 1977; Takasaka and Smith 1971; Wever 1978). The basilar membrane

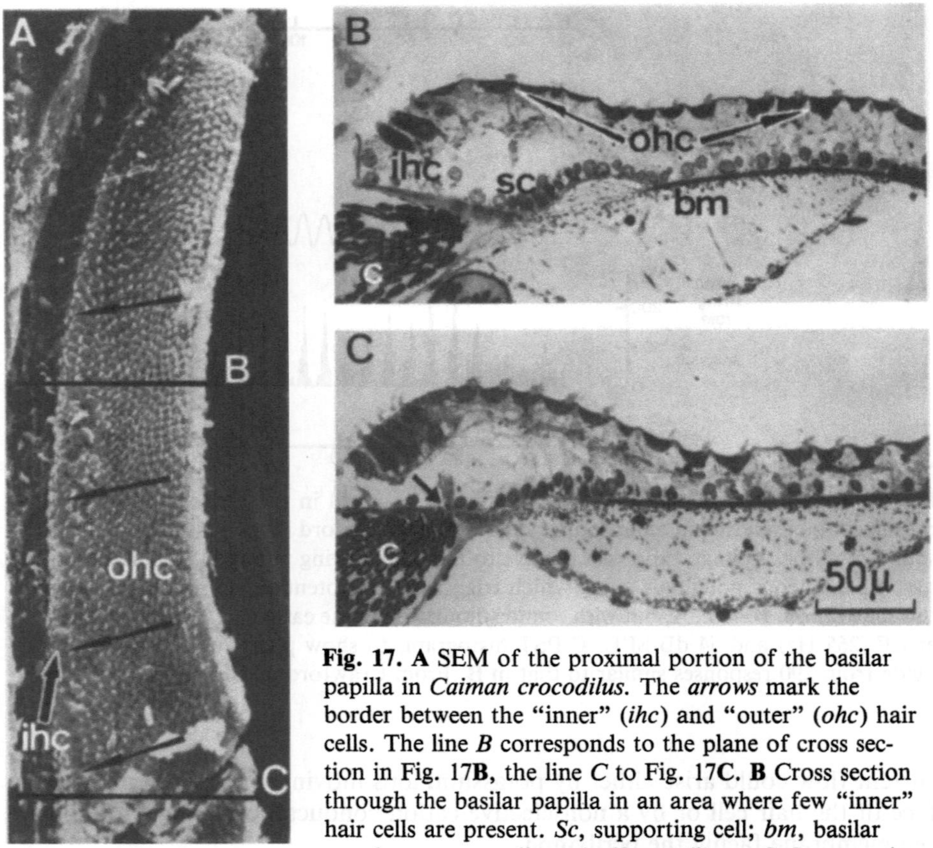

Fig. 17. A SEM of the proximal portion of the basilar papilla in *Caiman crocodilus.* The *arrows* mark the border between the "inner" (*ihc*) and "outer" (*ohc*) hair cells. The line *B* corresponds to the plane of cross section in Fig. 17**B**, the line *C* to Fig. 17**C**. **B** Cross section through the basilar papilla in an area where few "inner" hair cells are present. *Sc*, supporting cell; *bm*, basilar membrane; *c*, myelinated afferent fibers. **C** Cross section through the basilar papilla in an area with more "inner" hair cells. As in **B**, the full extent of "outer" hair cells is not shown. *Arrow,* habenula perforata. Magnification × 300. From von Düring et al. (1974) **A**, × 160; **B, C**, × 300

Fig. 18. SEM of the border between "inner" and "outer" hair cells of *Caiman.* The hair cells lie about 2 mm from the lagena and are seen approximately from above. From Leake (1977)

is long (4 mm) and curved. Two populations of morphologically distinct hair cells are seen, corresponding in their relative positions to the inner and outer hair cells of the mammal cochlea. The hair cells all show the same cilial orientation, but are not, as is the case in mammals, organized into separated parallel rows, but rather in a mosaic pattern within their respective regions. The populations are adjacent to each other, but apparently separately innervated. The basilar membrane is quite wide and at the lagenar end the hair cells can be seen lying up to 30 across in one cross section. There are about 8500 "outer" and 3000 "inner" hair cells (Figs. 17, 18).

Many of the hair cells on the inner side rest on the limbic structure, as in birds, rather than on the basilar membrane. As yet, the physiological data provide no functional correlation of these differences in the location, morphology and innervation of the hair cells. The inner hair cells are taller and closer together. The outer hair cells are flatter, with the cilia concentrated near the lateral margins of the hair cell surface (Fig. 18). All cilia seem to be surrounded by a honeycomb-like thick tectorial membrane which may be attached to the microvilli of the supporting cell surfaces. Hair cell bundles of inner hair cells are always taller than those of outer hair cells in the same region of the basilar papilla. In the direction of the lagena, the stereocilia of both kinds of hair cells gradually increase in height. Some evidence exists that the kinocilium may regress with age, as in mammals.

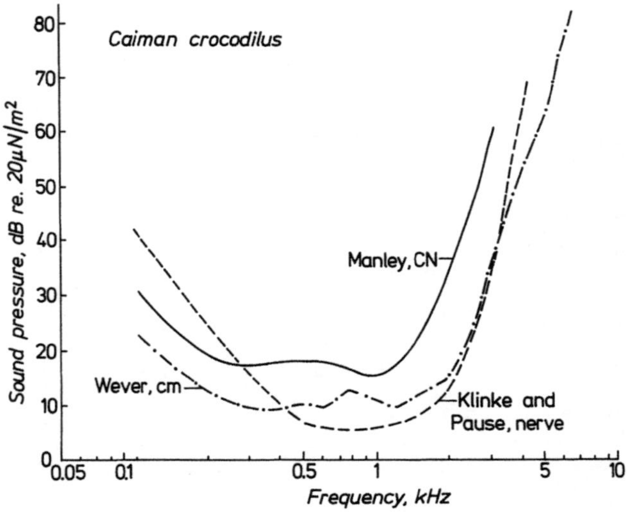

Fig. 19. Comparison of threshold audibility curve derived from the responses of auditory nerve fibers and neurons in the cochlear nucleus (*CN*) of *Caiman crocodilus* with a cochlear microphonic (*CM*) sensitivity function. Data from Klinke and Pause (1979), Manley (1970a and unpublished data), and Wever (1978)

In Fig. 19 are shown data from physiological measurements of the cochlear microphonic by Wever (1978), of single auditory nerve fibers (Klinke and Pause 1979), and of auditory units in the cochlear nuclei (CN) of the spectacled caiman, *Caiman crocodilus*. The frequency range covered is quite small (100 Hz – 5 kHz and, as is usually the case, the CM data overestimate the animal's high frequency sensitivity as well as provide no absolute threshold data (that is, the level of CM chosen for the measurements in arbitrary). There is, however, a general agreement between the two sets of data, as tends to be the case when such data are compared for animals with only unidirectionally oriented hair cells. When the hair cells are not all so oriented, large discrepancies between CM and neural data are evident (see Sect. 4).

In *Caiman*, Klinke and Pause (1979) have studied the properties of 390 auditory nerve fibers. These fibers were tuned between CF 30 and 2.8 kHz, with their Q values ranging from 1.5 to 7.0. Thresholds were as low as 5 dB sound pressure level (SPL), but varied up to 50 dB in any one animal. The tuning curves were more sharply tuned on the high frequency side (slopes of 30 – 180 dB/octave) than on the low frequency side (15 – 150 dB/octave), with only a few cases of symmetrical tuning curves. This is a curious and interesting difference from nerve fibers from the very similar bird cochlea, where an equally large proportion of fibers has a steeper low frequency slope (Manley and Leppelsack 1977).

In response to click stimuli, units fired repeatedly, as already noted for turtles, producing pronounced peaks in the peri-stimulus time histograms (PSTHs) computed from the responses. As in mammals, the first-peak latencies were strongly correlated with fiber CF. The number of peaks in the PSTH was correlated with the sharpness of tuning. Responding to clicks, 40% of units showed shorter latencies to a condensation click, as also seen in the turtle for all units (Paton et al. 1976). Twenty-two units tested with two-tone stimuli (one tone held steady at the CF, the other swept in frequency) produced in all cases two-tone suppression effects similar to those found in mammalian primary nerve fibers, with the low-frequency suppressive area being broader than the high-frequency area.

The nerve in *Caiman* was found to be tonotopically organized, which correlates with the previously demonstrated organization in the CN (Manley 1970a). Single pure-tone responses were tonic in nature, producing PSTHs as seen in the CN (Manley 1974) and corresponding to the "primary-like" category defined by Pfeiffer (1966). At higher sound intensities, the strongest response shifted to lower frequencies (a shift of up to 10% of CF per 10 dB, Fengler et al. 1978); thus the units discharge at higher rates to frequencies below CF than to CF itself (cf. Fig. 41). The CF of the fibers was also temperature dependent. This effect is discussed in Sect. 8.

3.4 Lizards

The inner ear of lizards shows great structural diversity, varying systematically both between and within the different families, but is "...sufficiently stable at the family level to be diagnostically characteristic of any one family" (Miller 1966a, 1974, 1978a). Some generalizations are possible for the most important families. Further details are available in Miller (1978a) and Wever (1978). Miller (1980) groups the different families of lizards as follows, according to relationships indicated by similarities in inner ear structure:

1. Iguanidae, Agamidae, Anguidae, Xenosauridae, Anniellidae
2. Chameleonidae (relationships unclear)
3. Lacertidae (uncertain if related more to skinks or to teiids)
4. Teiidae (possibly related to varanids)
5. Varanidae
6. Helodermatidae } possibly related to varanids
7. Lanthanotidae }
8. Scincidae, Feylinidae, Cordylidae, Xantusidae (closely similar in structural details)
9. Dibamidae and Anelytropsidae (relationships uncertain)
10. Gekkonidae and Pygopodidae (close affinities).

This listing resembles that of Wever (1978), although Wever lists a superfamily Varanoidea (Varanidae, Helodermatidae) and a superfamily Lacertoidea (Teiidae and Lacertidae). These groupings, based on inner ear structure alone, strengthen many previous ideas about relationships, derived from details of the skeleton, etc.

A summary of the typical hair cell orientation patterns and papilla shapes is shown in Fig. 20, from the work of Miller (1980). The shape of the papilla varies considerably, even being divided into two in some cases. There is almost always an area of unidirectional, abneurally oriented (that is, away from the neural limbus), less densely packed hair cells and a generally larger bidirectionally oriented area of more densely packed hair cells (Miller 1974). The supporting cells of lizards are reduced in size at their luminal ends (Bagger-Sjöbäck 1976; Miller 1978a) (see. Fig. 7). Lizards have two types of hair cells (Bagger-Sjöbäck and Wersäll 1973; Miller 1973a,b, 1974). Firstly, there are hair cells with short- to medium-long cilia, having the kinocilial bulb attached to the five longest stereocilia. Secondly, there are long-ciliated hair cells with a short, unattached kino-

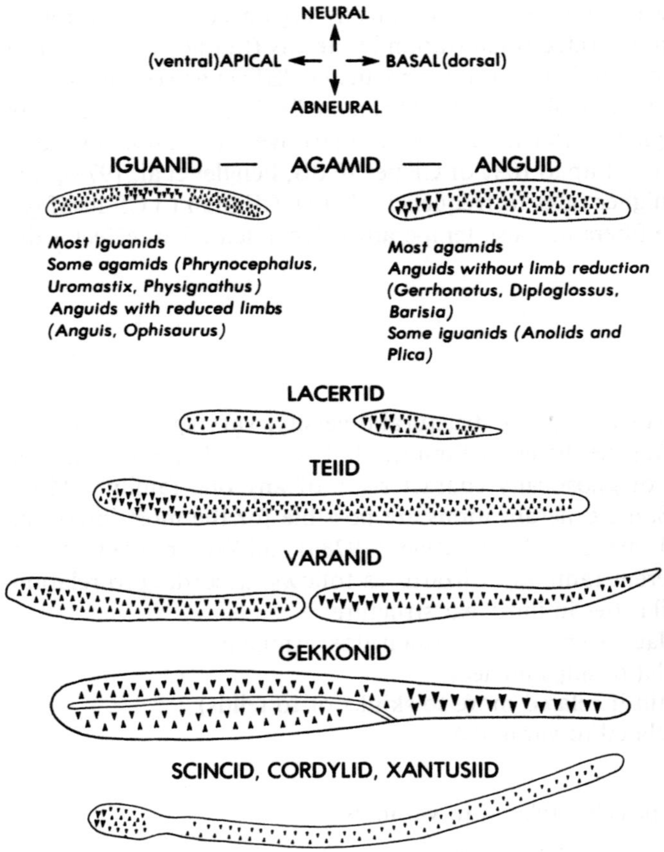

Fig. 20. Kinocilial orientation patterns in lizard papillae basilares. Typical shape of the basilar papilla and hair-cell orientation patterns in eight families of lizards. *Larger arrowheads* indicate areas of unidirectionally oriented hair cells. Not drawn to scale. From Miller (1980)

cilial bulb. The lizard basilar membrane, unlike that of caiman, turtles, birds, and mammals, is thickened with limbus-like material (Miller 1978b).

The three families Iguanidae, Agamidae, and Anguidae have papillae which, for the most part, consist of bidirectionally oriented, long-ciliated hair cell groups. They are also the only lizards with such freestanding long stereocilia which are not connected to a tectorial membrane as such. The other hair cells are short ciliated, unidirectionally oriented, and are covered by a limbus-attached tectorial membrane (Miller 1978a). This unidirectional area is either central or apical in location on the papilla. The pattern of kinocilial orientation in the bidirectional area of the lizard basilar papilla also shows family variations, some (such as geckos) showing strict orientation toward the midpapillary axis, others (such as Teiids and Varanids) randomly showing orientation toward the neural or abneural directions (Miller 1978a).

There are three larger studies of primary auditory nerve fiber responses in lizards, from the monitor lizard *Varanus bengalensis* (Manley 1977), the gecko *Gekko gecko* (Eatock et al. 1981; Eatock and Manley, 1981) and from the alligator lizard *Gerrhonotus multicarinatus* (Weiss et al. 1976). These data show a variety which matches the variation in the anatomy. In fact the three species almost cover the extremes of anatomy to be found. It is therefore reasonable to expect to be able to draw some preliminary conclusions about the function of the lizard basilar papilla.

3.4.1 The Alligator Lizard

In the alligator lizard *Gerrhonotus multicarinatus,* Weiss et al. (1976, 1978b) have studied basilar membrane motion, basilar papilla cell responses and auditory nerve fiber activity. The basilar papilla in the alligator lizard is short (0.4 mm

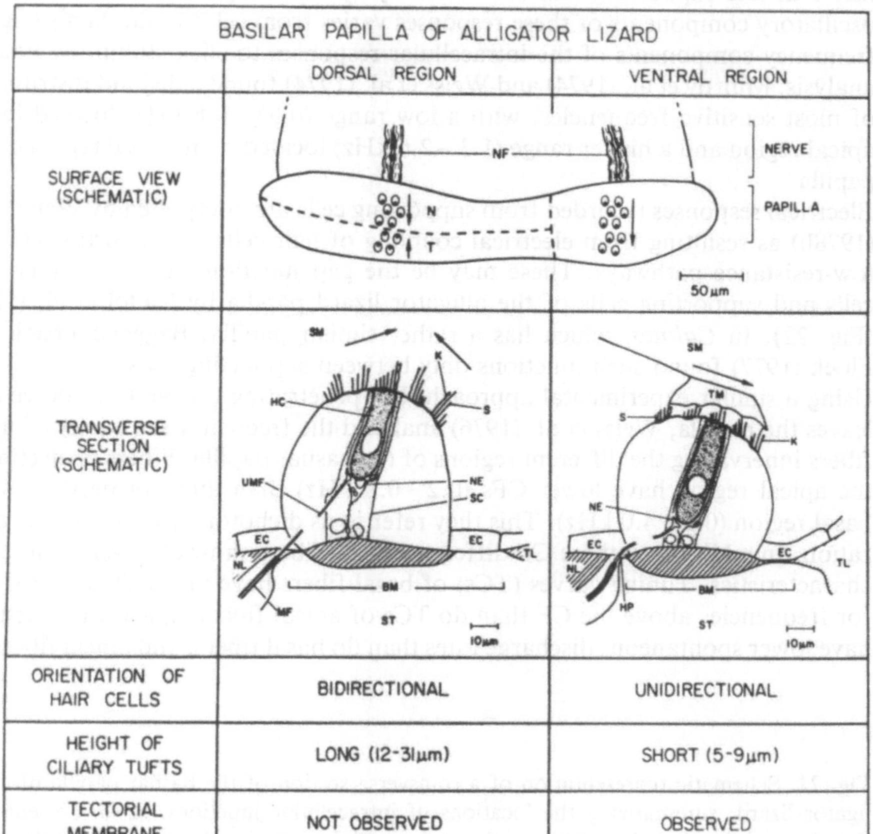

Fig. 21. The basilar papilla of the alligator lizard. The upper section shows a surface view of a right basilar papilla seen from the scala media, with a few hair cells and their orientations indicated on the neural limbus side (*N*) and triangular limbus side (*T*). Some of the nerve fiber (*NF*) bundles are also shown. In each of the transverse sections shown in the center, from the apical (ventral) and basal (dorsal) regions of the papilla, is shown one of the hair cells (Hc) with its adjacent supporting cells (*SC*) and nerve endings (*NE*). *UMF*, unmyelinated nerve fiber; *MF*, myelinated nerve fiber; *HP*, habenula perforata; *SM*, scala media; *BM*, basilar membrane; *ST*, scala tympani; *EC*, epithelial cell; *NL*, neural limbus; *TL*, triangular limbus; *K*, kinocilium; *S*, stereocilia; *TM*, tectorial membrane. The table summarizes the principal differences in morphology of the two regions of the papilla. From Weiss et al. (1974)

long), has only about 150 hair cells, and is divided into two adjacent but morphologically distinct regions. In the shorter apical region there are about 50 hair cells with short abneurally oriented cilia which contact a tectorial membrane (Fig. 21). In the basal region the hair cells are polarized in two opposing directions and the cilia are longer, the longest cilia on these hair cells being about 12 μ, increasing apically to about 31 μ. There is no tectorial membrane in the basal region. Afferent nerve terminals contact all hair cells, apparent efferent terminals only contact the apical hair cells (Weiss et al. 1978b).

Using glass micropipettes, intracellular responses to sound were recorded in both hair cells and supporting cells (identified by dye marking). The frequency of the oscillatory components of these responses varies from cell to cell. Analyzing the frequency components of the intracellular responses to click stimuli by Fourier analysis, Mulroy et al. (1974) and Weiss et al. (1974) found a disjoint distribution of most sensitive frequencies, with a low range (0.35 – 0.8 kHz) located in the apical region and a higher range (1.3 – 2.6 kHz) located in the basal region of the papilla.

Electrical responses recorded from supporting cells are interpreted by Weiss et al. (1978b) as resulting from electrical coupling of hair cells to supporting cells by low-resistance pathways. These may be the gap junctions found between hair cells and supporting cells of the alligator lizard papilla by Nadol et al. (1976) (Fig. 22). In *Calotes,* which has a rather similar papilla, Bagger-Sjöbäck and Flock (1977) found such junctions only between supporting cells.

Using a similar experimental approach, but penetrating the auditory nerve as it leaves the papilla, Weiss et al. (1976) analyzed the frequency sensitivity of nerve fibers innervating the different regions of the basilar papilla. Fibers projecting to the apical region have lower CFs (0.2 – 0.8 kHz) than those projecting to the basal region (0.9 – 4.0 kHz). This they refer to as dichotomous tototopic organization. In addition to these CF differences, the fibers show divergences in other characteristics. Tuning curves (TCs) of basal fibers have much shallower slopes for frequencies above the CF than do TCs of apical fibers, apical fibers tend to have lower spontaneous discharge rates than do basal fibers, and apical fibers ex-

Fig. 22. Schematic representation of a transverse section of the basilar papilla of an alligator lizard, summarizing the locations of intracellular junctions. *A – F* are enlarged drawings of the circled regions. *A*, the regions of the junctional complex near the endolymphatic ends of adjacent hair cells and supporting cells showing that tight junctions (zonulae occludentes, *ZO*) are always present between adjacent hair cells and supporting cells and between adjacent supporting cells. Desmosomes (zonulae and maculae adherens, *ZA* and *MA*) are also present within the junctional complex. Just below the complex, small gap junctions occur between hair cells and supporting cells, and larger gap junctions between supporting cells. *B*, extensive gap junctions occur between adjacent supporting cells. *C*, gap junctions, but not tight junctions, are found at the bases of supporting cells. *D*, simple squamous epithelium covering the perilymphatic surface of the basilar membrane. *E*, the endolymphatic regions of adjacent hyaline cells possess junctional complexes like those between supporting cells, and also large gaps junctions. *F*, hyaline and supporting cell junctions possess similar junctional complexes and gap junctions. From Nadol et al. (1976)

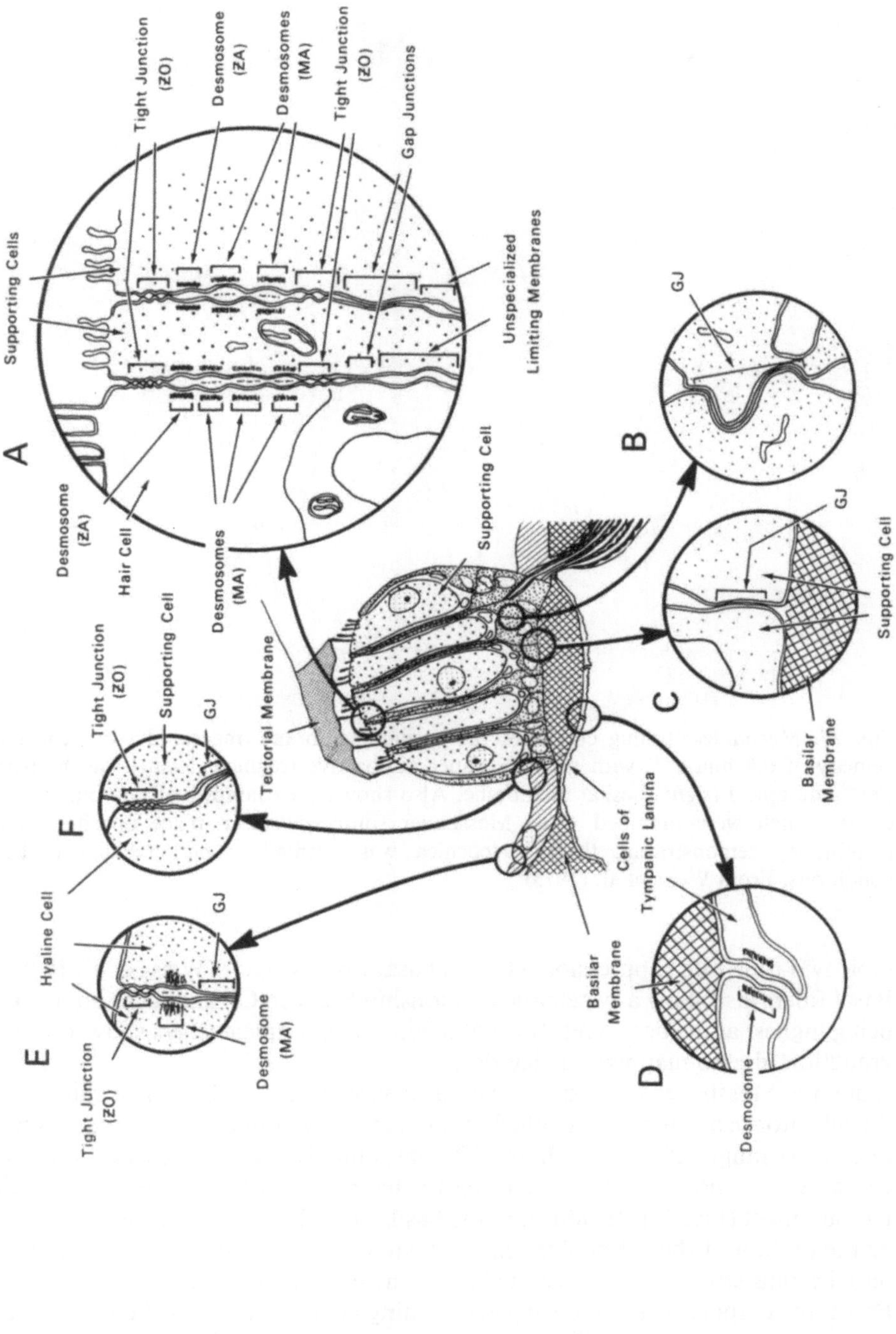
A
Supporting Cells
Tight Junction (ZO)
Desmosome (ZA)
Desmosomes (MA)
Tight Junction (ZO)
Gap Junctions
Unspecialized Limiting Membranes
Desmosome (ZA)
Hair Cell
Desmosomes (MA)
Supporting Cell
Tectorial Membrane
B
GJ
C
GJ
Supporting Cell
Basilar Membrane
F
Tight Junction (ZO)
Supporting Cell
GJ
E
Hyaline Cell
GJ
Tight Junction (ZO)
Desmosome (MA)
Cells of Tympanic Lamina
Basilar Membrane
D
Desmosome

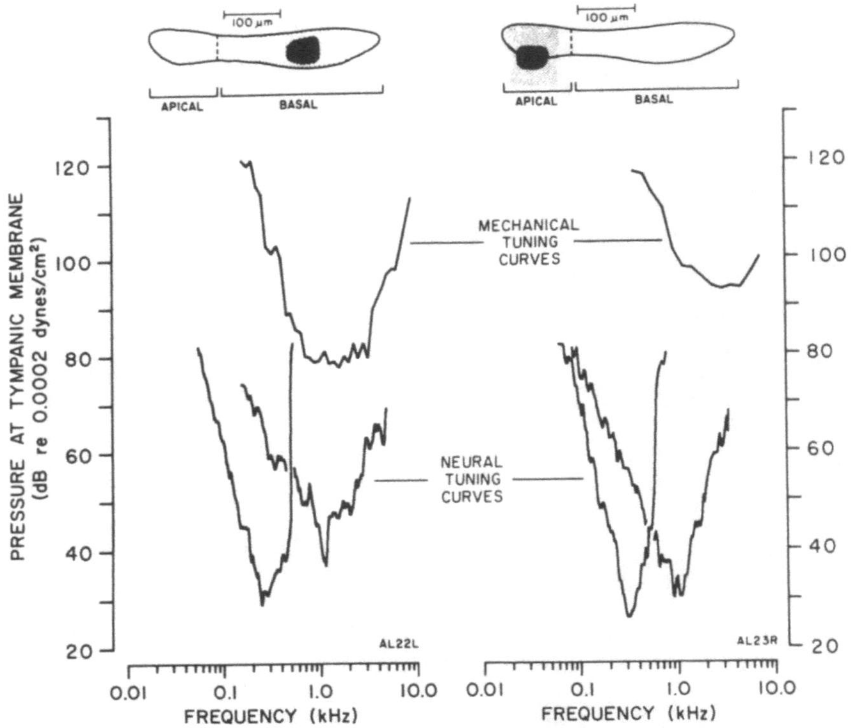

Fig. 23. Mechanical tuning curves of the basilar membrane (measured for a constant velocity of 0.8 mm s^{-1}) with the Mössbauer radioactive fragment located on the basal (*left*) and apical (*right*) basilar membrane. Also shown are examples of neuronal tuning curves which were obtained while Mössbauer sources were in place on the basilar membrane, demonstrating that the cochlea was normally functional under these conditions. From Weiss et al. (1978)

hibit two-tone rate suppression whereas basal fibers do not (Weiss et al. 1978b). Basal fibers also show a systematic relationship between CF and location, the CF being highest at the basal end. It is uncertain whether apical fibers show a similar gradation, due to mapping difficulties.

Using the Mössbauer technique, it was also shown that the basilar membrane of the alligator lizard is broadly tuned in the frequency domain, with best sensitivities in the range 1.5 – 4.1. kHz in different animals and the mechanical tuning characteristics not changing according to the location of the source during the measurement (Fig. 23). In addition, the basilar membrane motion closely followed the motion of the extracolumella of the middle ear, whose frequency characteristics thus strongly influence the basilar membrane motion.

From this it appears that the frequency tuning characteristics of the nerve fibers of the eighth nerve of the alligator lizard are not the result of any special fine tuning of the basilar membrane, but are, on the other hand, probably fully developed in the hair cells themselves. The nerve fibers may thus reflect a selectivity determined at a very peripheral level, as appears now to the case in mammalian nerve fibers (Russell and Sellick 1977). Because the basilar membrane of the alligator lizard shows no gradation in its width or, indeed, in its mechanical tuning, Weiss et al. (1978) suggest that the striking differences in the morphology of tectorial and ciliary structures result in different fine mechanical tuning at the ciliary level. The systematic variation in the length of the basal cilia may be responsible for the graded tonotopic organization of this region of the papilla.

3.4.2 The Monitor Lizard

The papilla of the monitor lizard, like that of the alligator lizard, is divided into two regions (Miller 1978a; Wever 1978). In *Varanus bengalensis,* these two regions are separated by a complete constriction of the papilla, a short gap in which not hair cells are found. The papilla is quite long (1.6 mm) and contains about 1800 hair cells (Fig. 24). The basal region is large and is itself divided into two adjacent regions. An apical area of about one-third of its length contains only unidirectional, abneurally oriented hair cells. The rest of the papilla in both the remaining basal and in the short apical region contains hair cells with a poorly organized, bidirectional orientation pattern. A tectorial membrane is present throughout.

Recording from the proximal nerve trunk and from nerve fibers near the basilar papilla, Manley (1977) established the presence of two groups of fibers and mapped these fiber responses onto the papilla. There is a larger group of fibers which have lower CFs (0.2 – 1.0 kHz) which are arranged in a tonotopic organization and innervate the basal region of the papilla. A smaller group of fibers responds to higher frequencies (CFs 1.3 – 2.8 kHz) and emanates from the apical region. Due to the problems of mapping a small area, the question of a regular arrangement of frequencies in the apical region is unsettled (Fig. 25).

Apical fibers have tuning curves which are on average less sharply tuned than those of basal fibers (Figs. 26, 27). Taking all fibers together, no particular asymmetry of the tuning curves was apparent. Fibers with tuning curves steeper on the high frequency side were as common as those steeper on the low side. No change in tuning characteristics was attributable to the transition from unidirectional to bidirectional hair cell orientation. Click response data, although not as clear as equivalent data in mammals, indicate a shift in the latency of the first spike upon reversal of click polarity in low CF fibers (up to 600 Hz), which emanate from the unidirectionally oriented hair cell area.

The fact that higher CF fibers did not was interpreted by Manley (1976) as indicating that hair cells of both polarities in these regions are innervated by the same nerve fibers. In addition, the PSTH discharge pattern in response to CF pure tones at 20 dB above threshold varied systematically according to CF, from a primary-like pattern (Pfeiffer 1966) for low CF units, through an intermediate type to a pattern with a lower discharge rate but a conspicuous initial peak or peaks for high CF units (Fig. 28). Phase-locking to the cycles of the stimulating tone occurred in low CF units, at least up to 900 Hz CF. Two-tone suppression was observed in the monitor lizard nerve fibers, as were occasional on-off responses (see below), but these phenomena were not studied systematically (Manley, unpublished data).

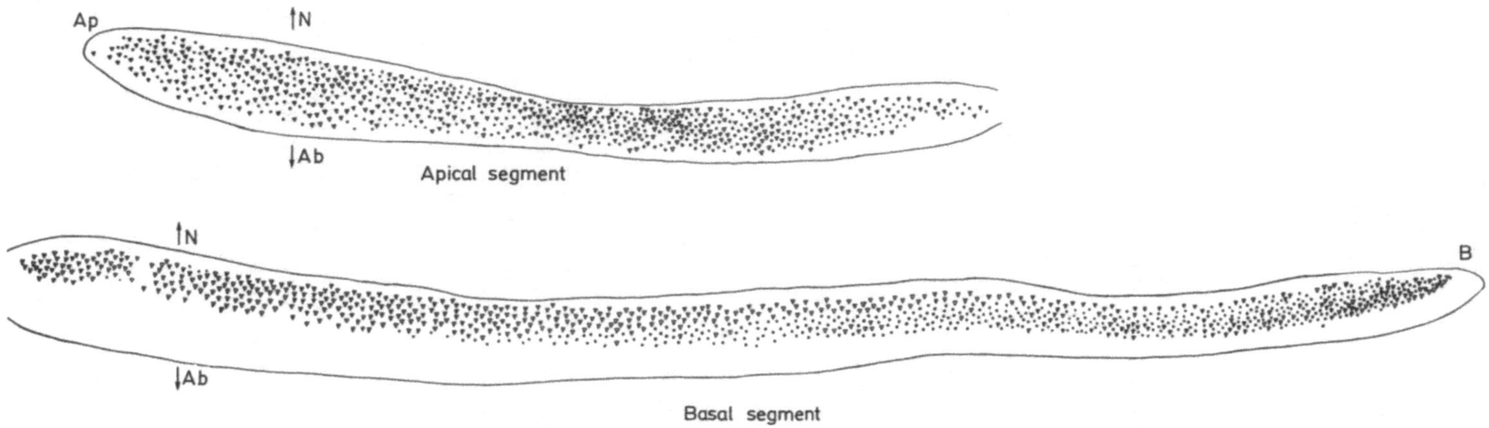

Fig. 24. Scale drawing of the basilar papilla of *Varanus bengalensis* (cf. Fig. 20), showing the orientation of each hair cell in both apical and basal segments. *Triangles* represent abneural orientation (*Ab*); *black dots,* neural (*N*) orientation. Occasional *open circles* represent cells whose orientation was not determined. From Miller (1978a)

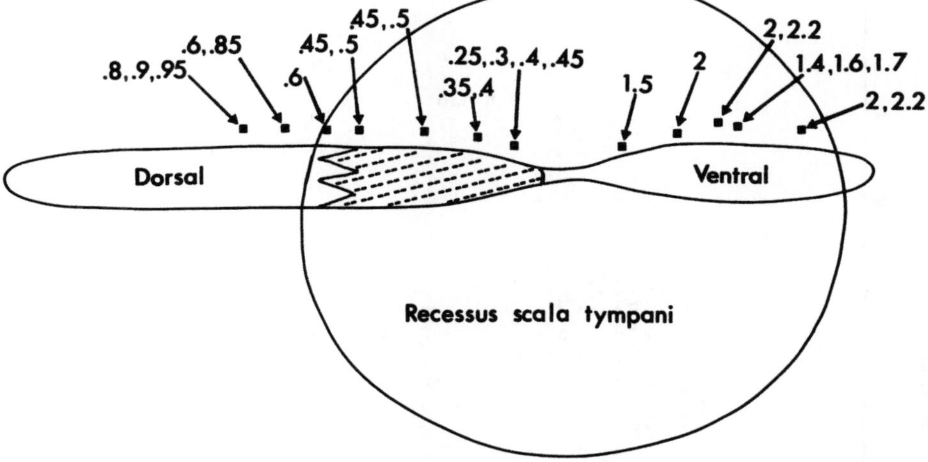

Fig. 25. Basilar papilla of *Varanus bengalensis* showing the area visible though the opening of the *recessus scala tympani. Dots* on the neural limbus indicate ink spots deposited by the recording electrode. The adjacent numbers are the CFs in Kilohertz, of units recorded in each location. The data are a composite from four animals. The two most dorsal ink spots represent estimates as to electrode location. Physiological data from these recordings suggested that the *hatched area* contained unidirectional hair cell orientation (cf. Fig. 24). From Manley (1977)

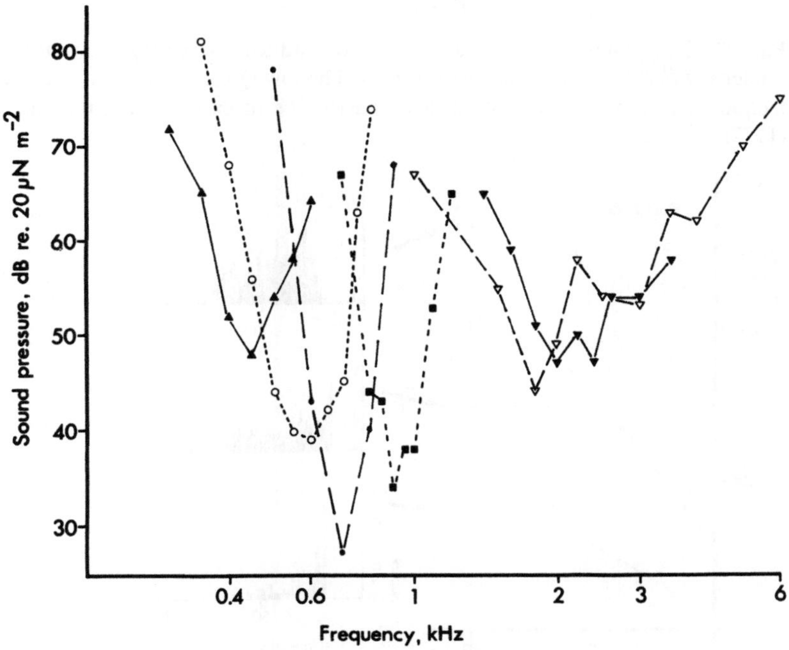

Fig. 26. Representative tuning curves from primary auditory neurons in *V. bengalensis.* Four units in the low CF group and two units in the high CF group are shown, not all from the same animal. The high frequency slopes of the high-CF group may be somewhat shallower than normal due to the opening of the scala tympani. From Manley (1977)

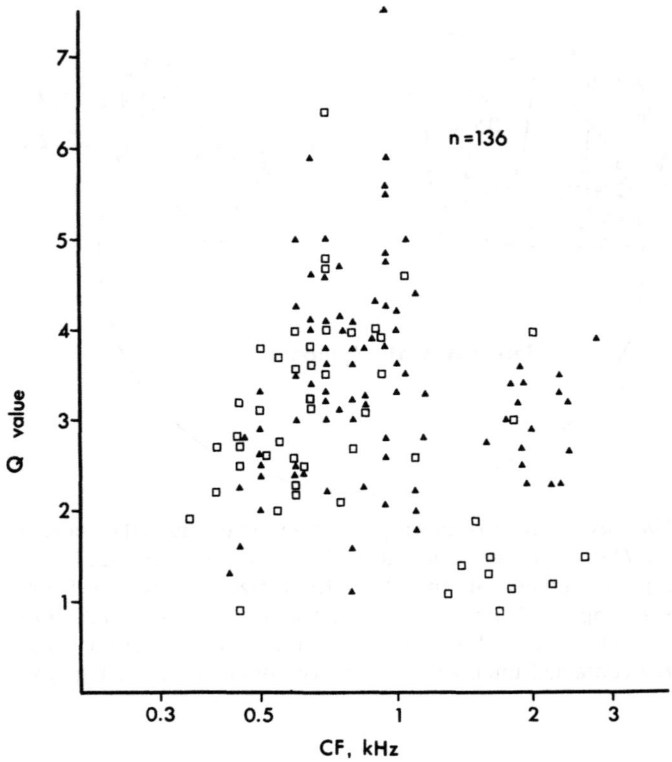

Fig. 27. $Q_{10\ dB}$ values plotted against CF for auditory nerve (*open squares*) and cochlear nucleus (*filled triangles*) units in *Varanus*. The low Q values for primary fibers in the high frequency group are probably at least partly due to exposing the papilla. From Manley (1977)

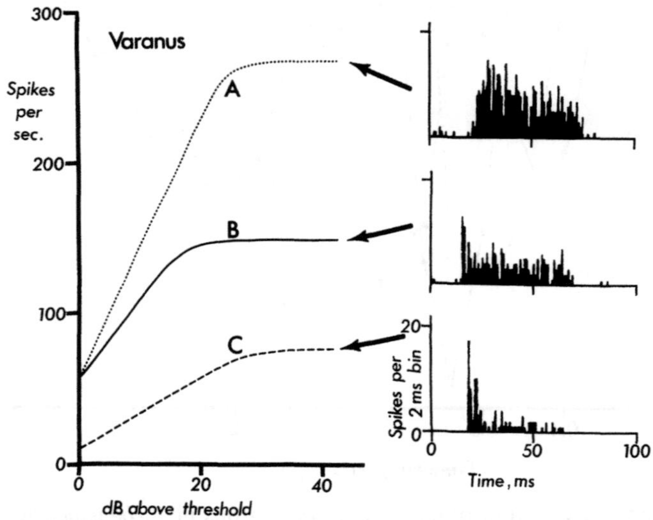

Fig. 28. Intensity functions (*left*) and PSTH patterns (*right*) of units of different CF (*A*, 650 Hz; *B*, 950 Hz; *C*, 950 Hz) to 50 CF tones at 25 dB above threshold intensity. Units are from the CN of *Varanus bengalensis*, but parallel data were obtained from the nerve. The filled (*A*), intermediate (*B*), and peaked (*C*) PSTH patterns are associated with different intensity function slopes and discharge maxima. From Manley (1976)

3.4.3 The Tokay Gecko

In the Tokay gecko, the auditory papilla reaches one of its highest states of development in the lizards. It is about 2.0 mm long and contains 2100 hair cells (Miller 1973a). The papilla shows a clear taper, from 130 μ apically to 45μ basally. At the basal end, it is covered for about one-third of its length by unidirectionally oriented hair cells. The rest of the papilla is covered by highly organized rows of bidirectionally oriented hair cell groups, in a doubly "bidirectional" pattern (Fig. 24; Miller 1973a, 1978a). Some kind of tectorial membrane covers virtually all hair cells, although the outermost rows of hair cells of the bidirectional area are attached only to freestanding tectorial masses, called "sallets" by Wever (1978). The innervation pattern in this species is not yet described.

Recordings from the nerve trunk have been reported by Eatock (1978), Eatock and Manley (1981), and Eatock et al. (1981). The auditory nerve units do not fall into two distinguishable populations on the basis of their frequency tuning properties, but rather show a continuum of characteristics. The TCs are quite sharply tuned, with a trend toward sharper tuning at higher CFs (CF range was 0.15 – 5 kHz). Using the standard Q measure for sharpness of tuning, a range up to 13 was found, higher than equivalent measures for mammals (Fig. 29). Figure 30A shows that, on average, the high frequency slopes of the TCs were steeper than the low frequency slopes, but Fig. 30B shows that this is mainly due to the majority of neurons of CF above 700 Hz having a steeper high frequency slope.

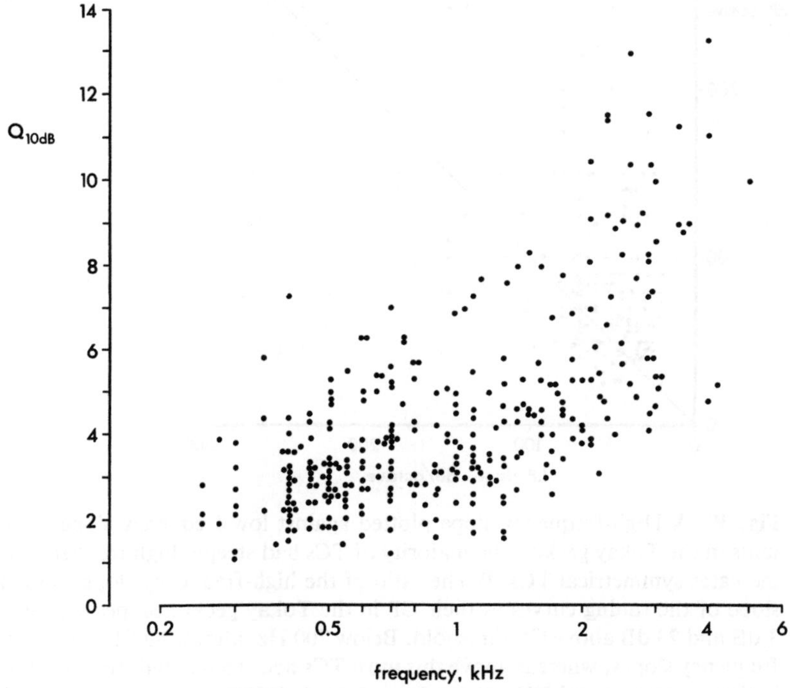

Fig. 29. $Q_{10\ dB}$ tuning curve sharpness-of-tuning coefficient for units in the auditory nerve of the Tokay gecko. From Eatock et al. (1981)

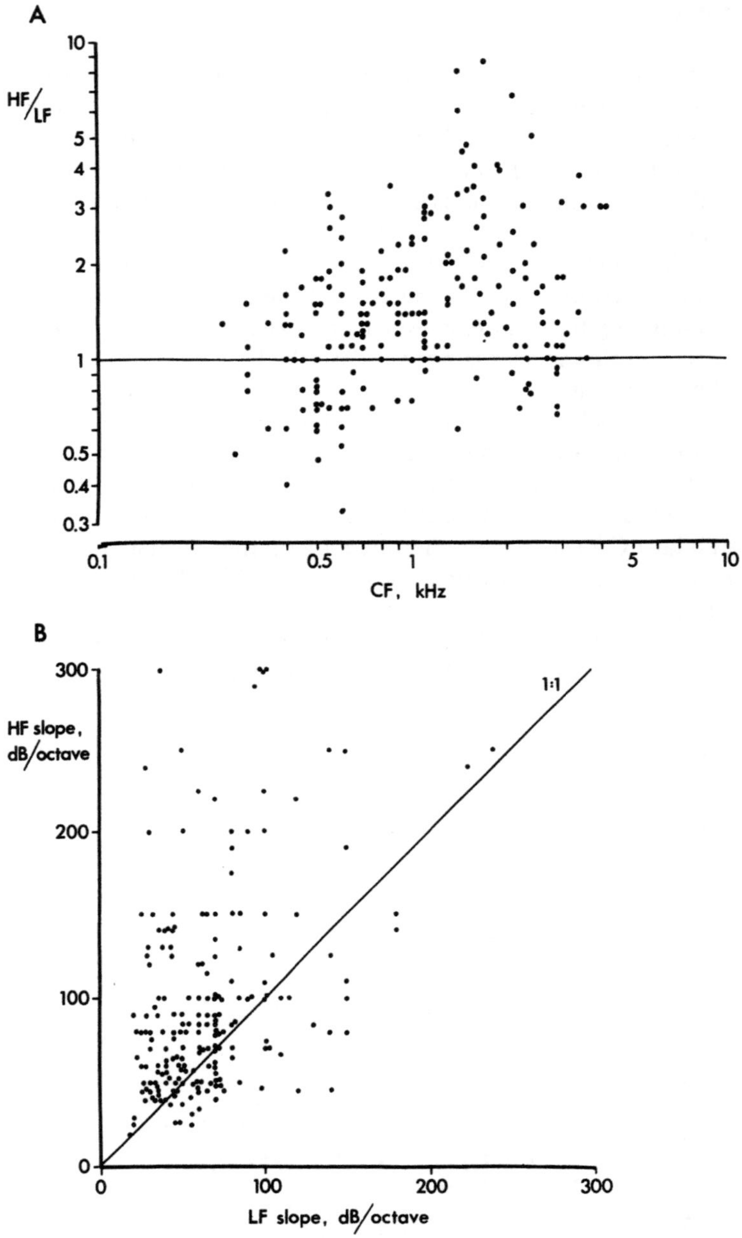

Fig. 30. A High-frequency slope plotted against low-frequency slope for auditory nerve units in the Tokay gecko. The majority of TCs had steeper high frequency slopes. The *line* indicates symmetrical TCs. **B** The ratio of the high-frequency slope to the low-frequency slope of the tuning curves vs their CF in the Tokay gecko. Slopes are measured between 3 dB and 23 dB above CF threshold. Below 700 Hz, almost half the units had steeper low-frequency slopes, whereas above this most TCs had steeper high frequency slopes. The *line* indicates symmetrical TCs. From Eatock et al. (1981)

Correcting these slopes for the middle-ear velocity response makes little difference in the whole picture. CFs were not randomly distributed in the nerve, a tonotopic organization being evident. Low CF fibers were encountered in the posterior part of the nerve. In the middle of the nerve, lower CFs (0.2 kHz) were found ventrally and higher CFs (up to 2.5 kHz) dorsally. CFs from 3 to 5 kHz were only found in the dorsal, anterior part the nerve, i.e., around the midline of the whole eighth nerve. In response to clicks, some units responded with shorter latency to a condensation click than to a rarefaction click, but insufficient data were collected to make definitive statements. This aspect needs further study.

For a number of fibers, PSTHs were constructed using pure-tone stimuli, and the discharge patterns at CF at 20 dB above threshold were similar to those obtained in the monitor lizard. Primary-like PSTHs were only obtained for units with CFs up to 0.7 kHz, whereas PSTHs with clear, brief onset peaks were obtained for units with CF greater than 0.7 kHz. Intermediate PSTH patterns were obtained from some units with CF between 0.4 and 2 kHz (Fig. 31). A change in the rise-time of the tone envelope or a change in the sound pressure level of the stimuli had little effect on the shape of PSTHs. Discharge rates were usually higher in low CF units for a given intensity above threshold. Phase-locking to the stimulus waveform occurred in almost all units tested with CF up to 700 Hz, but in no units tested having higher CF.

As these experiments with gecko were carried out in an anechoic chamber using free-field stimulation, and recording through a hole in the skin and muscle of

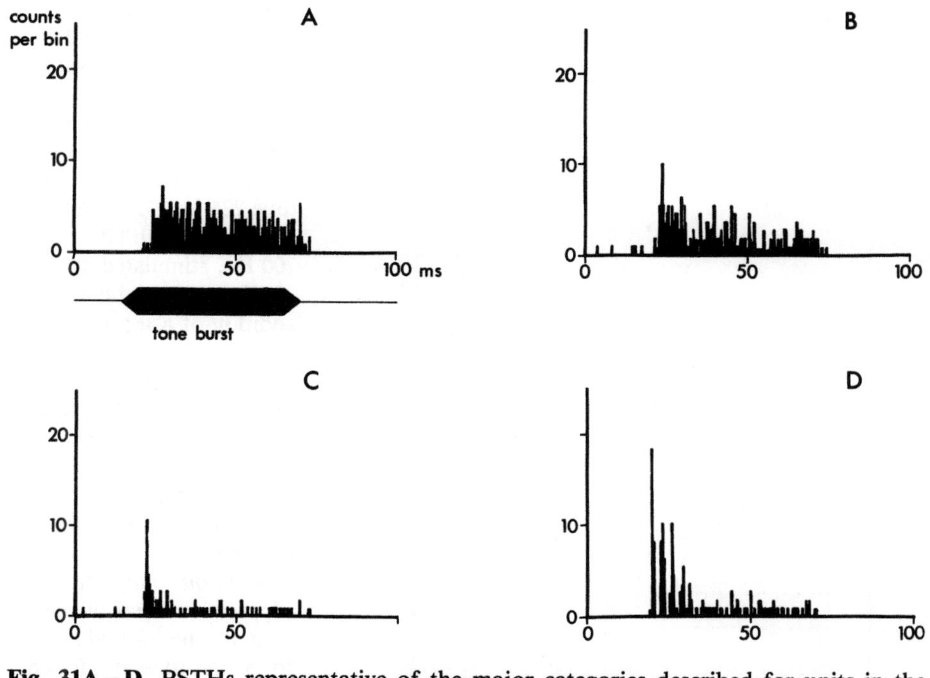

Fig. 31A–D. PSTHs representative of the major categories described for units in the auditory nerve of *Gekko gecko*. **A** Filled; **B** semipeaked; **C** single-peaked; **D** multiple peaked. In each case, the response is to 30 repeated 50-ms tone pulses at CF, 20 dB above threshold, rise and fall times 5 ms. From Eatock et al. (1981)

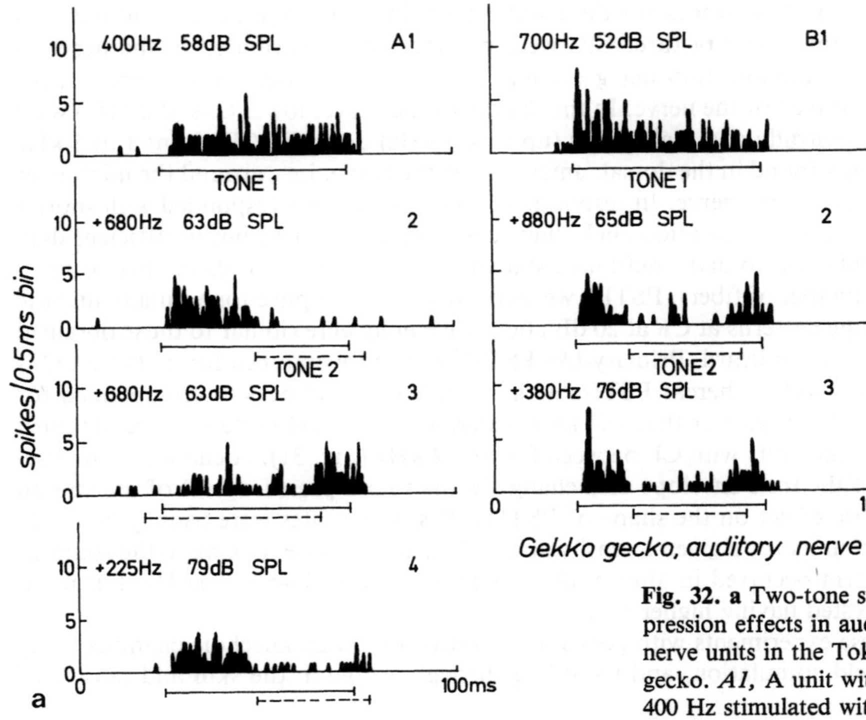

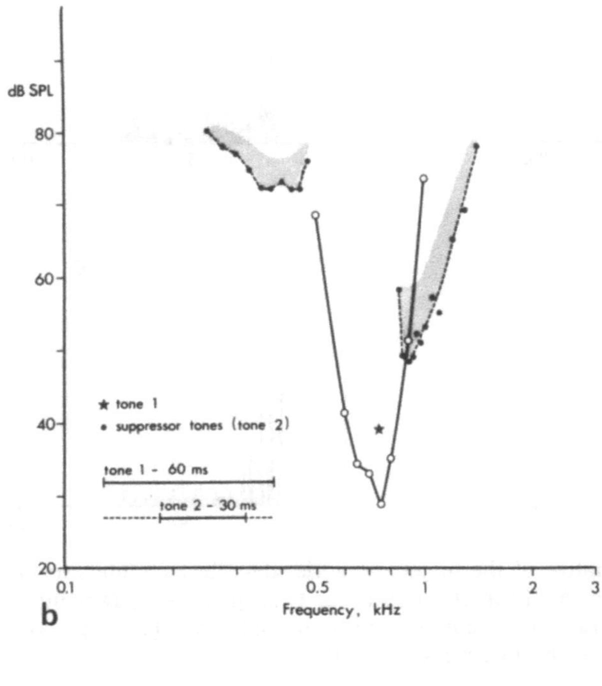

Fig. 32. a Two-tone suppression effects in auditory nerve units in the Tokay gecko. *A1,* A unit with CF 400 Hz stimulated with 20 repeated tones at CF, 15 dB above threshold (tone 1). *A2*, same, with a second tone (tone 2) added as indicated 680 Hz, 63 dB SPL. *A3*, second tone displaced in time. *A4*, second tone 225 Hz, 79 dB SPL. *B1*, A unit with CF 700 Hz, stimulated at CF, 20 dB over threshold. *B2,* addition of a second tone, 880 Hz, 65 dB SPL, as indicated. *B3* second tone 380 Hz, 76 dB SPL. Manley and Eatock, original figure. **b** Tuning curve of a single auditory nerve unit of the Tokay gecko showing excitatory threshold (*continuous line*) and suppression areas of the second tone (*dotted lines*) for a first tone at CF and 10 dB above threshold. Eatock et al. (1981)

the lower jaw, the possibility existed that the data were to some extent influenced by the sound field having access to the rear of the eardrum. In order to ascertain if significant effects were present, 48 units in the nerve of a gecko were stimulated using a closed acoustic system sealed over the skin around the eardrum (Pawson and Manley, unpublished data). The only difference observed was that the most sensitive units were about 10 dB more sensitive than in the free-field situation. Thus the opened mouth cavity may reduce the system's sensitivity by about 10 dB. The closed system confirmed the presence of what seems to be an important feature of the gecko data, and that is an obvious drop in sensitivity of the units centered near 1 kHz of about 10 – 15 dB (see Fig. 39). Also in this frequency region are found less steep tuning curve slopes, especially on the low-frequency side. As a consequence, high Q values are infrequent in units with CFs of 1 – 1.5 kHz. The above features are probably the consequence of changes in the hair cell and tectorial membrane patterns between different regions of the papilla. These changes are also manifest in the change in PSTH shape with increasing CF.

Although two-tone suppression was not systematically studied, it was observed in all primary units tested in *Gekko*. The CFs were between 0.4 and 2 kHz, and suppressive areas were present on both sides of the tuning curve (Fig. 32). The suppression was obtainable at lower sound pressures of the second tone when the second tone was of higher frequency, as observed in mammals (Sachs and Kiang 1968) and birds (Sachs et al. 1974; Manley and Leppelsack, unpublished data). In units with one shallow tuning curve slope, suppression was difficult or not possible to obtain using frequencies near this slope, as it was generally in insensitive units. This has also been shown for mammals (Robertson 1976).

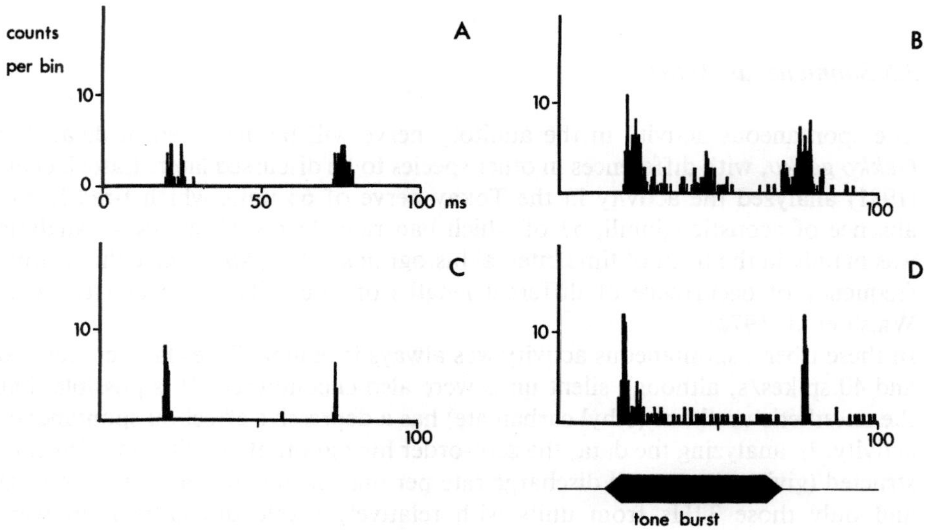

Fig. 33A – D. On-off PSTH patterns for some units in the auditory nerve of the Tokay gecko. **A** Response of a unit with CF 150 Hz to 30 repeated CF tones at 62 dB SPL. **B** As in a, 72 dB SPL. **C** A unit with CF 2.1 kHz (CF threshold 36 dB SPL) responding to 1.6 kHz at 70 dB SPL (1.6 kHz threshold = 69 dB SPL). **D** A unit with CF 300 Hz (CF threshold = 33 dB SPL) responding to 700 Hz at 89 dB SPL (700 Hz threshold 84 dB SPL). Tone burst 50 ms. Eatock et al. (1981)

In 10 of the 410 fibers studied by Eatock (1978) and Eatock et al. (1981), on-off discharge characteristics were observed, although no systematic search was undertaken (Fig. 33). Of 77 fibers which were the subject of PSTH analysis, 5 were on-off units. Two of these were on-off at all frequencies tested, and the other three were on-off only at frequencies removed from the CF. In all five cases, thresholds for such on-off discharge behavior were above 60 dB SPL. Similar on-off discharge patterns have been observed in avian auditory fibers by Gross an Anderson (1976) and Manley and Leppelsack (to be published).

3.4.4 Additional Data

In the skink *Trachysaurus rugosus,* a small amount of comparative data for the nerve fiber responses is available (Johnstone and Johnstone 1969). Unfortunately, data were only presented for four fibers in terms of the tuning curves, and the PSTH data are difficult to compare with those from other species. In addition, the papilla hair cell pattern is not described exactly in this species. The CF range given (0.7 – 3 kHz) is smaller than one would anticipate from a skink, but is probably due to the small sample size. The PSTH patterns shown in this work indicate a strong peak at the onset. This pattern would be expected from a skink papilla, with mainly bidirectional orientation of the hair cells (Miller 1974). Johnstone and Johnstone (personal communication) also found a seasonal variation in the magnitude of the summating potentials of the cochlea and in the activity of the auditory nerve fibers, a finding which deserves further, more detailed study.

3.5 Spontaneous Activity

The spontaneous activity in the auditory nerve will be discussed in detail for *Gekko gecko*, with differences in other species to be discussed later. Eatock et al. (1981) analyzed the activity in the Tokay nerve of 64 units which fired in the absence of acoustic stimuli, 53 of which had rates below 15 spikes/s. Analysis was mainly in the form of time interval histograms (TIHs) which give the relative frequency of occurrence of different lengths of interspike intervals (see, e.g., Walsh et al. 1972).

In these fibers, spontaneous activity was always irregular. Rates lay between 1.6 and 40 spikes/s, although silent units were also encountered. It is possible that the anesthetic urethane (ethyl carbamate) has a depressive effect on spontaneous activity. In analyzing the data, the zero-order histogram of activity was also constructed (giving the overall discharge rate per unit of time during the recording) and only those TIHs from units with relatively stable discharge rates were accepted. Six representative TIHs are shown in Fig. 34. TIHs from units with CFs above 0.5 kHz (e.g., Fig. 34A at 1 kHz, rate 29/s and Fig. 34B, at 1.05 kHz, rate 11/s) were generally comparable to TIHs from mammal auditory nerve units (Manley and Robertson 1976; Walsh et al. 1972) and birds (Manley and Leppelsack 1977) despite the lower rates. The distribution of intervals is quasi Poisson. Units with CF below 0.5 kHz, on the other hand, produced TIHs which

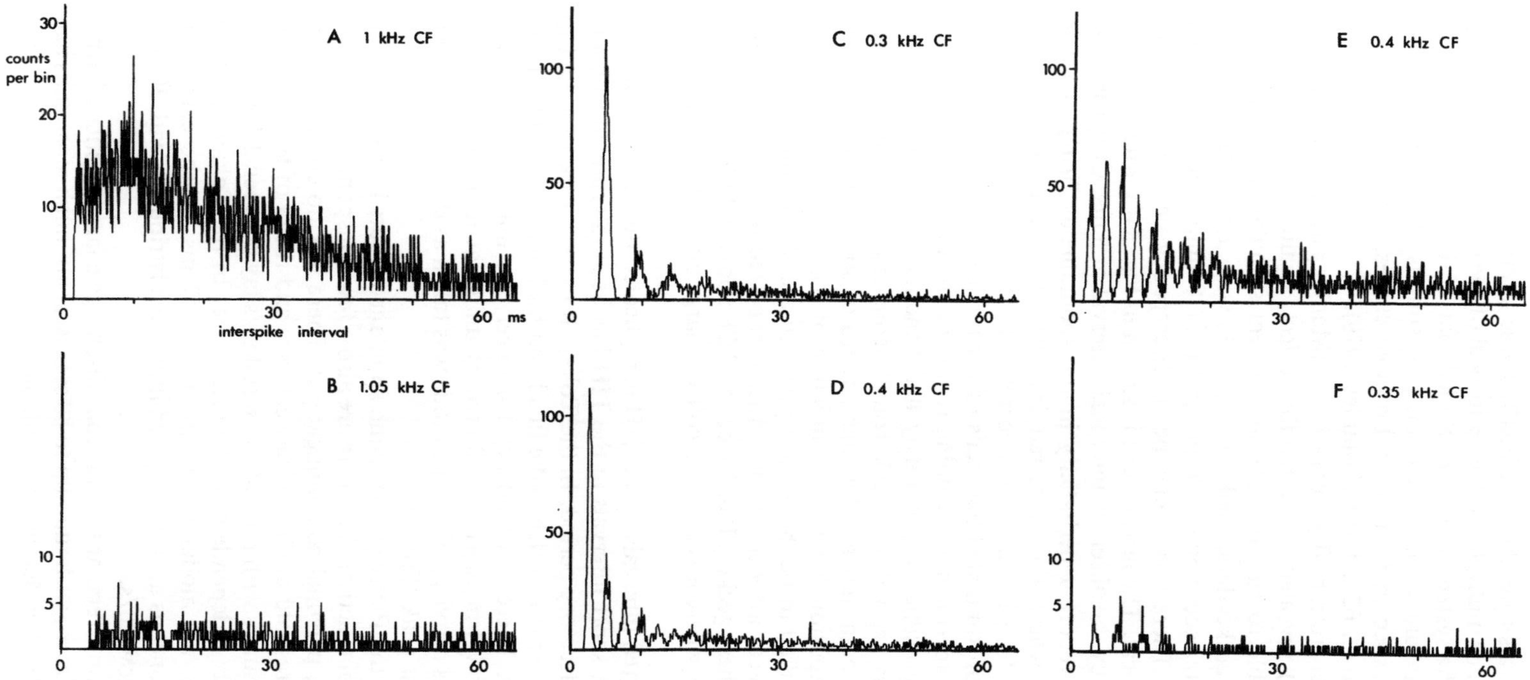

Fig. 34A – F. Representative TIHs of spontaneous activity in six auditory nerve fibers of the Tokay gecko. *Abscissa,* interspike interval (milliseconds) resolution 0.1 ms bins. *Ordinate,* relative frequency of interspike intervals of different lengths. Ordinate scales vary. In each case, unit CF is indicated. **C – F** display preferred interval lengths. From Eatock et al. (1981)

showed preferred intervals (Fig. 34C, D, E, F). These intervals tended to correspond to the reciprocal, or multiples of the reciprocal of the units' CF. Following extensive checking of the system, it was concluded that, if these data were not in fact spontaneous, these units were responding to sounds at a sensitivity far greater than could be expected from their threshold curves, for they often had quite high thresholds (e.g., 65 dB SPL) (Manley 1979).

Analysis of similar data from bird primary fibers (Manley and Leppelsack 1977 and unpublished data) revealed a comparable effect, again in circumstances where it seemed possible to rule out any inadvertent stimulation of the units (Manley 1979). In the gecko data, peaks in the TIHs were always seen in units with CF below 0.4 kHz, occasionally from 0.5 kHz CF units, but never from units with higher CF. There was sometimes a discrepancy between interpeak interval and CF reciprocal. The largest seen is shown in Fig. 34C, where a CF of 300 Hz is not matched by the observed interpeak interval of 5 ms. Crawford and Fettiplace (1980) also report a periodicity in the spontaneous firing of turtle auditory nerve units, where the preferred intervals also correspond to the reciprocal of the CF. This periodicity was correlated with the observed narrow-band noise present in unstimulated hair cells of the turtle's basilar papilla.

After several control experiments (including removing the middle ear to reduce the ear's sensitivity to possible room noise by 40 dB and looking at triggered correlograms between nerve spikes and background acoustic noise), these authors also conclude that the data indicate that this pattern of spontaneous activity is not a consequence of spurious acoustic stimulation of the ear.

The mode of TIH is taken to be the highest point on the plot of intervals (Fig. 35). The exponential decay after the mode, which is expected for primary auditory fibers, was seen in these gecko TIHs for units of CF above 0.4 kHz, especially easily in high rate units where it is easier to collect a large quantity of intervals in a short time.

The presence of preferred intervals in the TIH complicated the pattern. The low numbers or absence of short intervals in the TIH has been interpreted as being related to the unit's refractory period. In the Tokay, however, longer modes (up to 70 ms) have been observed, which make this hypothesis inadequate to explain the whole time period. There is a tendency for lower rate units to have longer modes. All modes below 5 ms were contributed by units having preferred intervals in the spontaneous activity, where the mode was taken as the interval corresponding to the highest peak (Fig. 34).

In the caiman, Klinke and Pause (1979) found spontaneous rate between 0.5 and 80 spikes/s, with irregular activity but there are also a few nonspontaneous fibers (Smolders and Klinke, personal communication). There was no correlation between rate and unit threshold at CF, or between rate and the Q of the unit. The modes of the TIHs for units with rates above 8 spikes/s lay between 2.5 and 6 ms and seemed to be fairly independent of the mean rate. In the Tokay gecko for units with these rates, the modes varied from 3 to 17 ms. In the caiman, it appears that the spontaneous activity may be bimodally distributed, with 30% of fibers having rates below 20/s.

Weiss et al. (1976) report irregular spontaneous discharge characteristics of nerve fibers in the alligator lizard, with rates up to 80 spikes/s, although most units had rates below 50/s. No further details are available in this species.

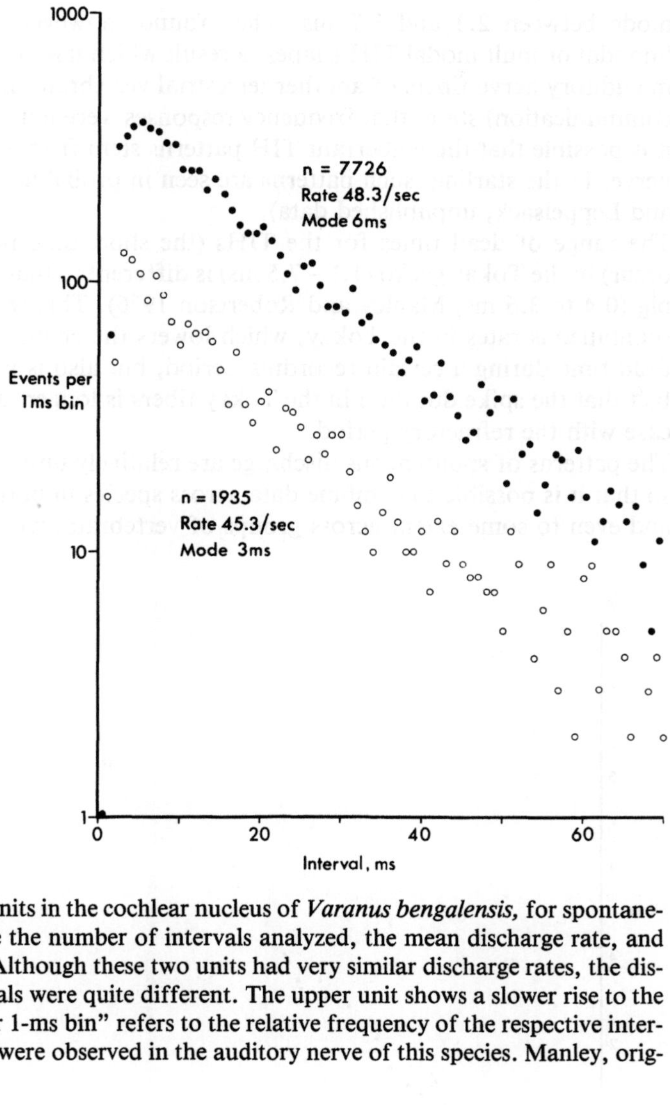

Fig. 35. TIHs from two units in the cochlear nucleus of *Varanus bengalensis,* for spontaneous activity. In each case the number of intervals analyzed, the mean discharge rate, and the mode are indicated. Although these two units had very similar discharge rates, the distributions of short intervals were quite different. The upper unit shows a slower rise to the mode (6 ms). "Events per 1-ms bin" refers to the relative frequency of the respective interval length. Similar TIHs were observed in the auditory nerve of this species. Manley, original figure

In the monitor lizard auditory nerve, Manley (1977) reported that spontaneous rates for 22 fibers were between 0.65 and 52 spikes/s and a mean rate of 14/s. TIHs generally had exponential decay patterns such as that shown in Fig. 35. Some units in *Varanus* in both CN (53 units) and auditory nerve showed a slow rise to the mode in the TIH, while others showed a more rapid rise to a more sharply defined mode (Fig. 35). Taking nerve and CN data together it was observed that units originating in the apical papilla had a higher mean rate (22/s) than those originating in the basal papilla (mean 10.3/s). This difference did not seem to have correlates in the other behavior of the units and needs further study. In *Trachysaurus rugosus,* the bobtail skink, Johnstone and Johnstone (1969) found the rates of spontaneous firing in 15 units to lie between 10 and 45 spikes/s. TIH for the activity of six of these fibers were obtained, each having a mode between 2.3 and 3.7 ms. They found, however, in some units stable bimodal or multimodal TIH shapes, a result which has never been demonstrated in auditory nerve fibers of another terrestrial vertebrate. B. Johnstone (personal communication) states that frequency responses were not studied in all fibers, so it is possible that these aberrant TIH patterns stem from the adjacent vestibular nerve. In the starling, such patterns are seen in probable lagenar fibers (Manley and Leppelsack, unpublished data).

The range of dead times for the TIHs (the short time intervals which do not occur) in the Tokay gecko (1.1 – 7.5 ms) is different to that reported in the guinea pig (0.4 to 3.6 ms, Manley and Robertson 1976). This reflects partly the lower spontaneous rates in the Tokay, which lowers the chance of obtaining the true dead time during a certain recording period, but also is probably related to the fact that the spike duration in the Tokay fibers is longer, as is probably also the case with the refractory period.

The patterns of spontaneous discharge are relatively uniform in their properties, so that it is possible to combine data across species in both CN and nerve units, and even to some extent across groups of vertebrates without major discrepan-

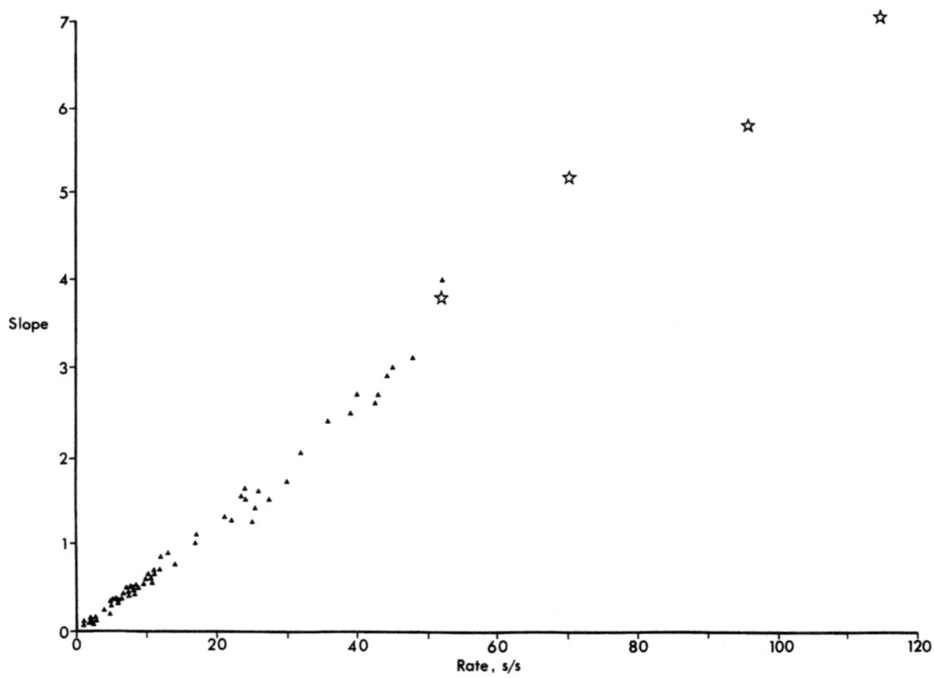

Fig. 36. Similarity of TIH plots for spontaneous activity between units from different species. The slope of the exponential decay of the TIH in *arbitrary units*, measured using identical plotting axes, is plotted against the unit's mean rate. *Triangles* represent individual units in the nerve of the monitor lizard and cochlear nuclei in the monitor lizard, caiman, and leopard lizard. The data from different species were not shown separately as the variations were so small. Four sample high-rate units (*stars*) are from the auditory nerve of the guinea pig (Manley and Robertson 1976), illustrating a continuity even with mammalian data

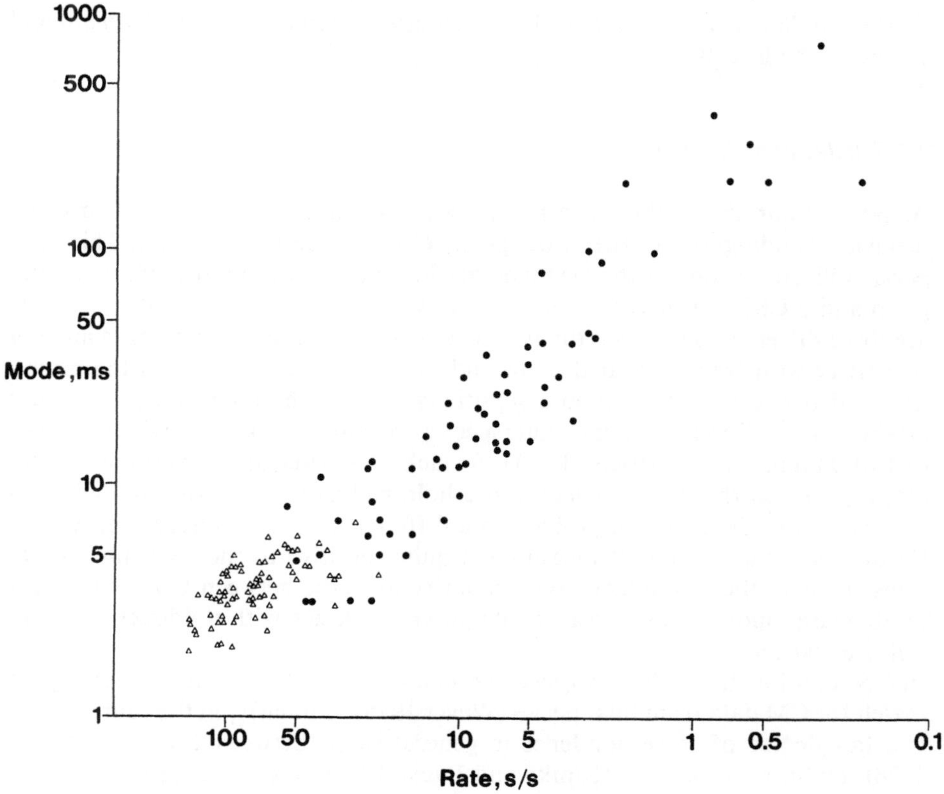

Fig. 37. Double logarithmic plot of TIH mode versus mean spontaneous rate. *Open triangles* represent higher-rate units in the auditory nerve of the guinea pig (Manley and Robertson, 1976); *filled circles,* units in the auditory nerve or cochlear nucleus of caiman, Tokay gecko, and monitor lizard. As in Fig. 36 different species are not separately indicated in the data (cf. Fig. 35). The product of the rate and mode represents a constant of about 250. Note that the *abscissa* is plotted with reversed scale

cies. This points to similarities in the underlying generator processes. It can be shown, for example, that the slope of the exponential portion of the TIH plot is closely related to the spontaneous rate in spikes per second (Fig. 36). Similarly, the mode of the TIH is directly related to the rate when studied over a very wide range of rates with, however, minor discrepancies between species (Fig. 37).

4 Cochlear Microphonic Data

Wever and his co-workers' data on the CM and cochlear anatomy of more than 200 species of reptiles have been recently reviewed in depth by Wever (1978). The data will thus be described here only insofar as it is relevant to the few species where neuronal recordings are available. They will be dealt with in two groups –

firstly, papillae with unidirectional hair cell orientation (turtles and *Caiman)* and secondly, the lizards.

4.1 Turtles and Caiman

As pointed out above, there is a fairly good agreement between CM data and a behavioral audiogram for the turtle species *Crysemys scripta* (Fig. 11). This comparison illustrates one of the best matches in shape between a behavioral audiogram and a CM "sensitivity" function. As Wever (1978, p.30) points out, there are three differences between the two sensitivity curves; in the rate of change of sensitivity with frequency at the low end of the frequency scale, in the prominence of the region of maximum sensitivity, and in the upper frequency limit attained. In addition, of course, interspecies comparisons of sensitivity by means of CM data are made difficult by (1) the lack of a criterion for defining threshold, (2) the fact that the true neural threshold and CM output vary in their relationship across frequencies, perhaps in a different way in different species, and (3) the fact that the output measured is quite sensitive to the recording conditions. Despite these problems, the similarity in shape between CM and behavioral/neural audiograms is reasonably good in species with unidirectional hair cell orientation.

In CN data for the turtle *Terrapene carolina* (Manley 1970b) can be seen a good match for CM data from this species (Wever 1978). Similarly, in the caiman there is a fair degree of correspondence in general shape between CM data (Wever 1978), auditory nerve data (Klinke and Pause 1979), and cochlear nucleus data (Manley 1970a) (Fig. 19). It seems therefore reasonable to assume that CM data can be used for comparisons between the species in this group in terms of their overall frequency response, although absolute sensitivity comparisons have no foundation.

4.2 Lizards

The wealth of data assembled on the CM of lizards cannot be effectively discussed here. I will therefore point out one major discrepancy when compared with the neural data available from a few species. Figs. 38 and 39 compare these sets of data in three well-investigated species – the monitor and alligator lizards and the Tokay gecko. It is clear that in each case, if we match the low frequency sensitivities, there is a very poor equivalence at the higher frequencies. It appears that this important difference can be traced to the presence in lizards of bidirectional hair cell orientation patterns. Because the electrical activity of such opposed hair cells will be more or less out of phase, a major part of their activity will not be represented in the recorded CM. If the responses from the two populations are not equal in magnitude and out of phase, of course, more complex relationships can be expected. However, it would appear to be unreasonable to expect a measurement of the fundamental component of CM to give comparable estimates of sensitivity or activity stemming from unidirectional areas on the one hand, and bidirectional areas on the other. From mapping data in the alligator and monitor

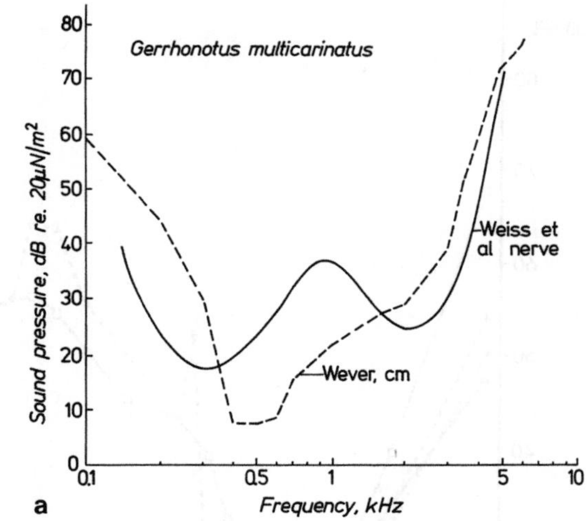

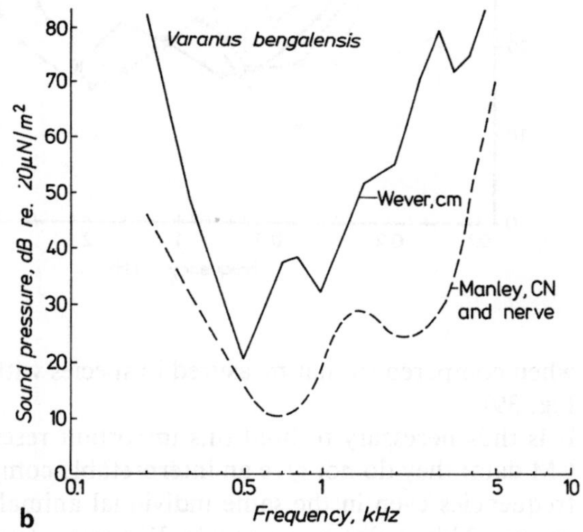

Fig. 38. a A comparison of CM sensitivity function (redrawn from Wever, 1978) and auditory nerve audibility function for the alligator lizard (estimated from the single-unit data of Weiss et al. 1976). **b** A comparison of a CM sensitivity function (redrawn from Wever 1978) and auditory nerve and CN audibility function for the monitor lizard (neuronal data from Manley 1976, 1977)

lizards, we know that the unidirectionally oriented areas contain hair cells which respond best to low frequencies and the bidirectional areas to higher frequencies. If this is true as a general rule in lizards, the mismatch at high frequencies would be explained: the CM there gives no reasonable estimate of sensitivity. That the unidirectional areas respond to low frequencies also receives support from Miller's (1975, and see Miller, 1978a, p. 391) study of the cochlear nuclei of lizards, where the size of the nucleus magnocellularis, which seems to be primarily involved in the analysis of low frequencies, is directly related to the development of the unidirectional area of the papilla. In addition, it can be seen by comparing Figs. 39 and 40 that, in species of gecko where the unidirectional areas are very small or absent, low-frequency CM sensitivity is exceedingly poor

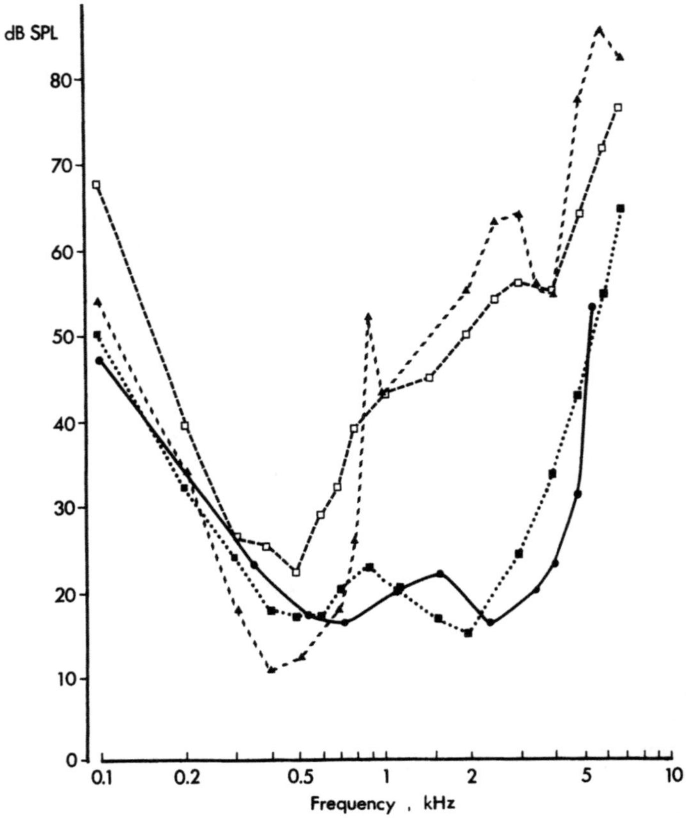

Fig. 39. A comparison of available audibility data for the Tokay gecko. Shown are CM sensitivity functions *open squares*, Hepp-Reymond and Palin (1968) *closed triangles*, Werner and Wever (1972), and audibility curves derived *filled squares* from single-unit data in the cochlear nucleus (Manley 1972c) and *filled circles* auditory nerve. From Eatock et al. (1981)

when compared to that measured in species with large unidirectional areas (e.g., Fig. 39).

It is thus necessary to hold this important reservation with regard to the lizard CM data: they do not give an interpretable comparison of sensitivity at different frequencies even in the same individual animal, much less between species and groups. This is also a reason why Wever's attempts to correlate hair cell population with CM output were not very successful (Wever 1978). Comparison of CM sensitivity between groups should be avoided, as there is no clear way to extend the statements to meaningful absolute threshold comparisons between species. Thus it has little meaning, and may be misleading to make such comparisons, as seen in Ridgeway et al. (1969), who state,

A comparison of this turtle with other animals in terms of cochlear potentials is of interest. In relation to the pigeon, *Chelonia mydas* is about equal in sensitivity up to 100 and at 600 Hz, but in the region between it is definitely superior. In relation to the cat, this turtle shows about the same degree of sensitivity in the low tones ...

Kauffmann (1974) and Kauffmann and Schwartzkopff (1971) report a study on the metabolic dependence of CM in the caiman. Using condensation and rarefaction clicks as stimuli, they studied these effects separately for positive and negative components of the CM (an upward movement of the basilar membrane

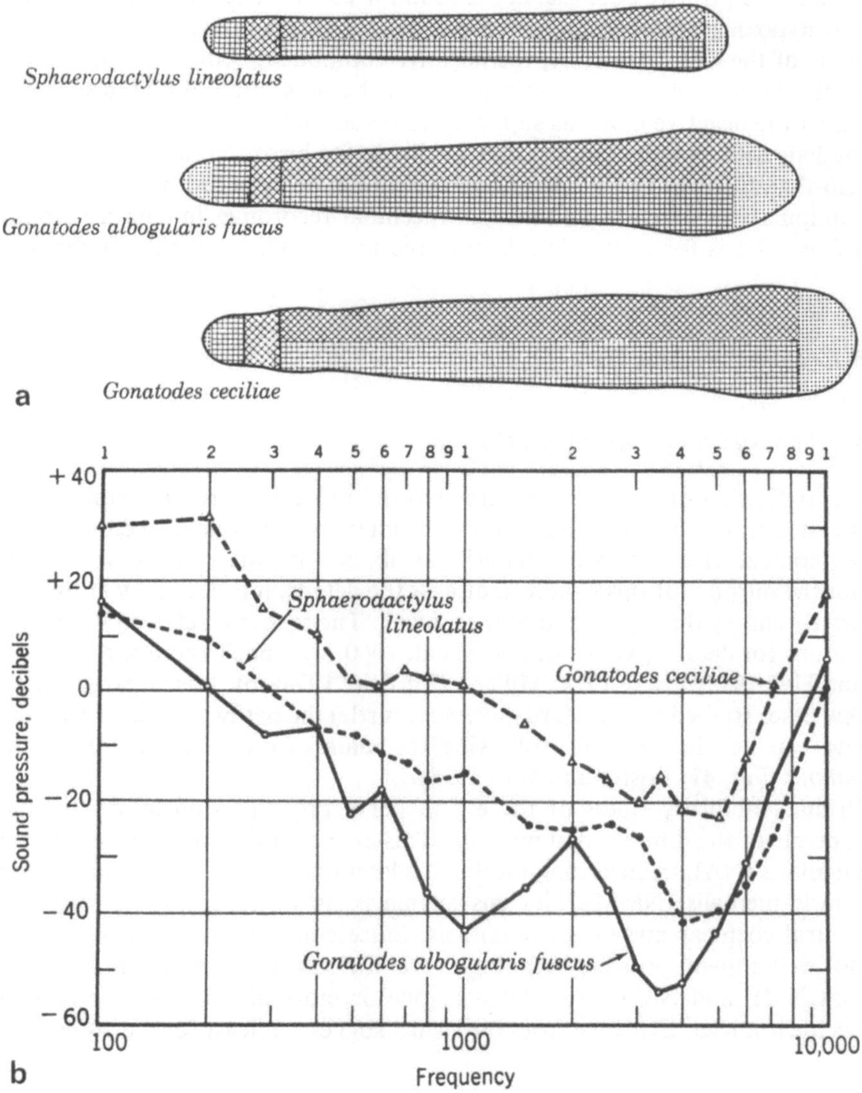

Fig. 40. a Schematic drawings illustrating the absence of unidirectionally oriented hair cell areas in three species of Sphaerodactylinid gecko. The *dotted areas* at the ends are free of hair cells; the two kinds of cross-hatching illustrate regions of bidirectional hair cell orientation either beneath sallets or beneath a regular tectorial membrane. Wever (1978). **b** CM sensitivity functions for the three species of geckos whose basilar papillae are illustrated in Fig. 40a. The absence of unidirectional hair cell orientation leads to a lack of low-frequency CM output (cf. Fig. 39; which shows CM data in *Gekko gecko*, which has a largish area of unidirectionally oriented hair cells). From Wever (1978)

produced a positive potential at the round window). These authors demonstrated that hypoxia, hypothermia, and metabolic inhibitors influenced the two components of the CM differently, the negative component being much more sensitive to interference. A model was proposed for the generation of CM in sauropsids in which the negative CM was supposed to be dependent on active ion transport in the hair cells. Similar findings are available for birds (Necker 1970) and the cat (Göttl and Klinke 1977). These findings should be extended by similar metabolic manipulations carried out during intracellular recordings in hair cells to provide a firmer basis for comparing them to recent neurophysiological findings.

5 Cochlear Nucleus Data

5.1 The Auditory Pathway in the Brain

The definition of the brain centers involved in audition results from anatomical investigations, including degeneration studies and physiological data, all taken in the context of comparative material available for other vertebrates. While it is not the purpose of this review to discuss the details, it is necessary at this point to briefly survey the reptile auditory pathway. The reader is referred to other publications for details (Ariens Kappers et al. 1960; Campbell and Boord 1974; Foster and Hall 1975; Leake 1974; Miller 1975; Pritz 1974a,b). It is apparent that in the few cases studied (some lizards, *caiman,* turtle) the pathways are basically similar and can be illustrated by following the main afferent connections in *Iguana iguana* (Fig. 41; Foster and Hall 1978).
Primary auditory fibers of the eighth nerve project to nuclei of the auditory tubercle of the dorsomedial medulla. This area contains three nuclei: the nucleus angularis (NA), nucleus magnocellularis lateralis (NML), and nucleus magnocellularis medialis (NMM). Nucleus angularis, in its connectivity, resembles the ventral cochlear nucleus of mammals. Since connectivity of the other nuclei is not well known, possible homologies are difficult to determine, but it is possible that NML and NMM are related to one or more divisions of the mammalian cochlear nuclei. Evidence from birds may suggest, at least for the crocodilia, that

Fig. 41. Summary of the ascending auditory pathways in the lizard *Iguana iguana*. The ▶ *heavy black line* indicates the primary ascending auditory pathway. *Smaller black lines* indicate less prominent auditory pathways. The *dashed lines with arrowheads* indicate suggested projections. The *dashed line* down the center indicates the midline. *A*, nucleus angularis; *Ai*, area intercollicularis; *aa*, ampulla, anterior semicircular canal; *ah*, ampulla, horizontal semicircular canal; *ap*, ampulla, posterior semicircular canal; *CG*, substantia grisea centralis; *Ctx*, cortex; *DVR*, dorsal ventricular ridge; *d*, torus semicircularis, nucleus dorsalis; *LL*, nucleus lemnisci lateralis; *M*, nucleus medialis; *ML*, nucleus magnocellularis lateralis; *MM*, nucleus magnocellularis medialis; *MS*, macula sacculi; *MU*, macula utriculi; *ml*, macula lagena; *OS*, nucleus olivaris superior; *Pb*, papilla basilaris; *Pn*, papilla neglecta; *St*, striatum; *TA*, nucleus corpus trapezoidum; *TSnc*, torus semicircularis, nucleus centralis; *TeO*, tectum opticum. From Foster and Hall (1978)

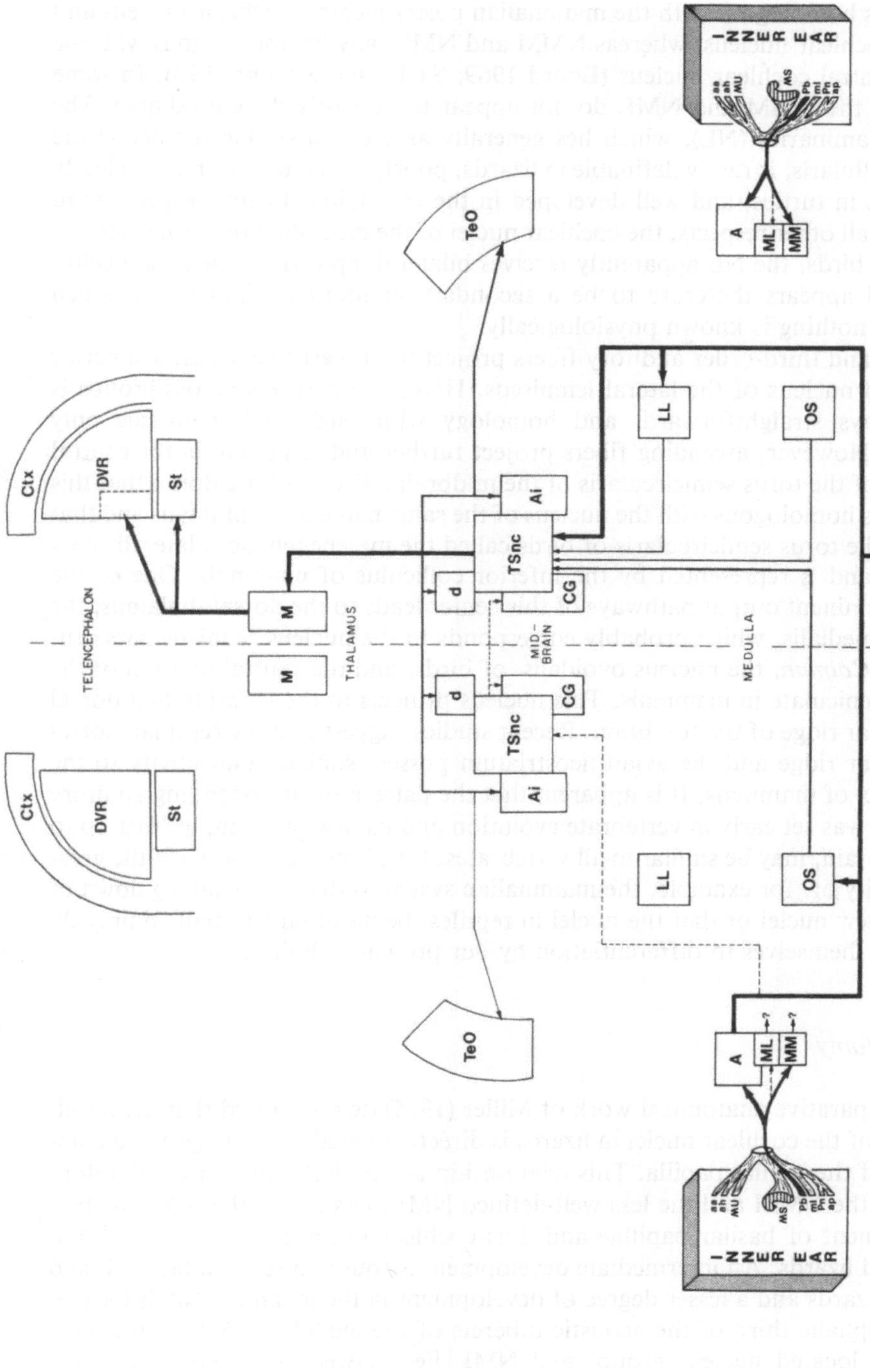
INNER EAR
A
ML
MM
TeO
LL
OS
TB
Ai
TSnc
CG
d
M
St
DVR
Ctx
TELENCEPHALON
THALAMUS
MID-BRAIN
MEDULLA
INNER EAR
A
ML
MM
TeO
LL
OS
TB
Ai
TSnc
CG
d
M
St
DVR
Ctx

the NA is homologous with the mammalian posteroventral cochlear nucleus and dorsal cochlear nucleus, whereas NMM and NML may be homologous with the anteroventral cochlear nucleus (Boord 1969; Sachs and Sinnott 1978). In some families, the NMM and NML do not appear to be easily distinguishable. The nucleus laminaries (NL), which lies generally as a sheetlike nucleus below the magnocellularis, is rarely definable in lizards, poorly developed in snakes, clearly definable in turtles, and well developed in the crocodilia. In this respect, as in virtually all other respects, the cochlear nuclei of the crocodilia resemble those of birds. In birds, the NL apparently receives bilateral input from the magnocellularis and appears therefore to be a secondary auditory nucleus about which virtually nothing is known physiologically.

Second- and third-order auditory fibers project to at least two areas, a superior olive and nucleus of the lateral lemniscus. Here, comparisons across groups is not always straightforward, and homology with birds and mammals only general. However, ascending fibers project further and terminate in the central nucleus of the torus semicircularis of the midbrain. There is little doubt that this nucleus is homologous with the nucleus of the same name of amphibians and that part of the torus semicircularis of birds called the mesencephalicus lateralis pars dorsalis and is represented by the inferior colliculus of mammals. One of the most prominent output pathways of this center leads to the dorsal thalamus, the nucleus medialis, which probably corresponds to the nucleus reuniens pars centralis of *Caiman,* the nucleus ovoidalis of birds, and the ventral division of the medial geniculate in mammals. This nucleus projects to the striatum and dorsal ventricular ridge of the forebrain. Recent studies suggest that the reptilian dorsal ventricular ridge and the avian neostriatum possess similar connectivity to the neocortex of mammals. It is apparent that the pattern of the ascending auditory pathway was set early in vertebrate evolution and its organization, at least up to the midbrain, may be similar in all vertebrates. It is, however, unclear if the great complexity of, for example, the mammalian system is due to the laying down of wholly new nuclei or that the nuclei in reptiles, being often so small, simply do not lend themselves to differentiation by our present techniques.

5.2 Anatomy

The comparative anatomical work of Miller (1975) demonstrated that the development of the cochlear nuclei in lizards is directly related to the length and complexity of the basilar papilla. This relationship is especially true for the development of the NMM and the less well-defined NML, less so for the NA. The best development of basilar papillae and thus cochlear nuclei is found in teiid and gekkonid lizards. An intermediate development is found in scincid, lacertid, and anguid lizards and a lesser degree of development in the iguanids. NA is located in the cephalic third of the acoustic tubercle of the medulla, NMM is the most caudally located nuclear group, and NML lies between the two nuclei as a sparsely cell-populated area (Fig. 42, Miller 1975).

Glatt (1975b) divided the reptiles into two groups, the "crocodile" types, with a clearly homologous NM and a nucleus laminaris (Crocodilia, Teiidae, Varanidae, Agamidae, Iguanidae, and Chameleonidae). In the "Lacerta" types (all

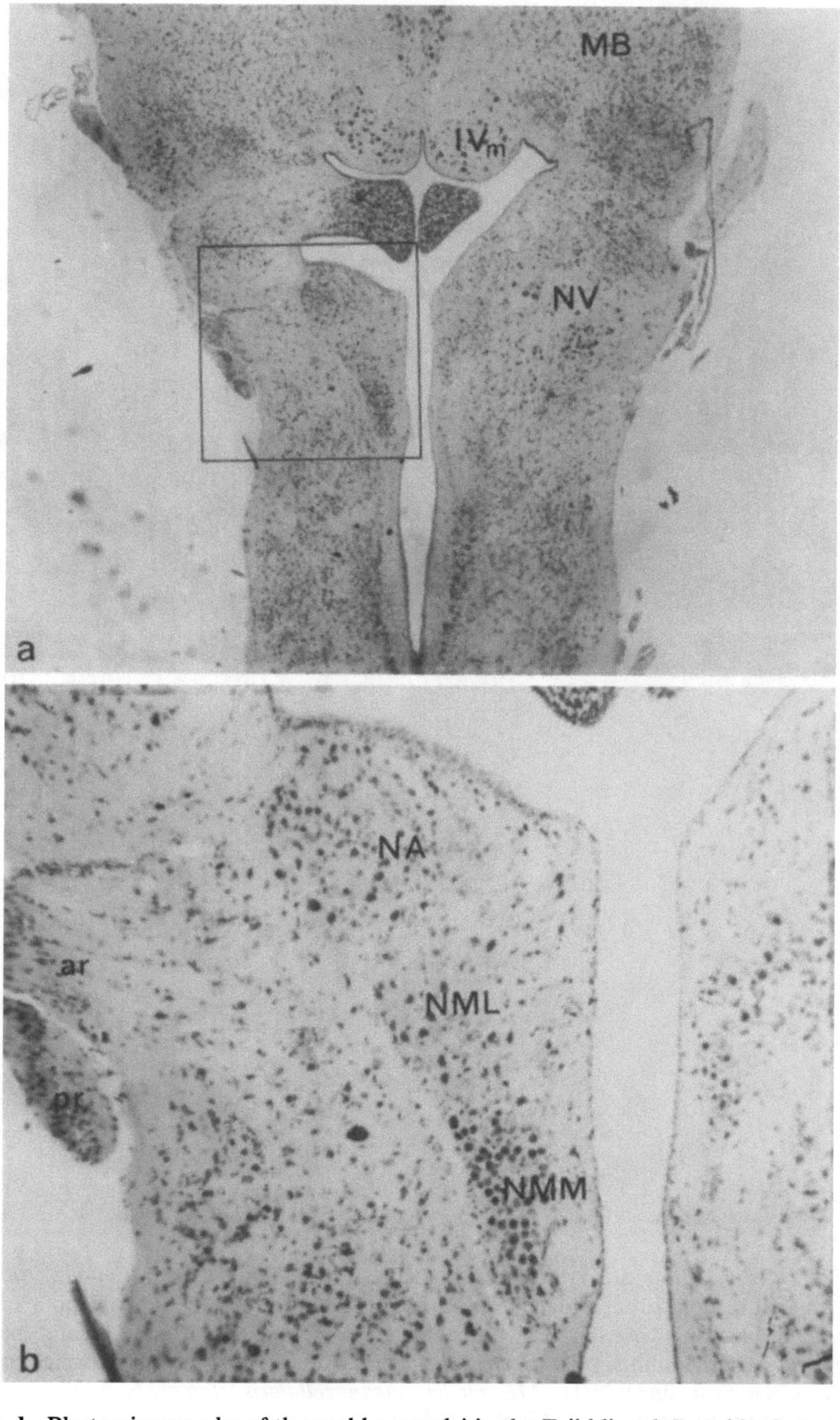

Fig. 42 a, b. Photomicrographs of the cochlear nuclei in the Teiid lizard Cnemidophorus tigris. **a** Frontal (horizontal) section showing the extent of the cochlear nuclei, which occupy the medial part of the area outlined. *MB*, midbrain, *IVm*, trochlear nerve, motor nucleus; *NV*, nucleus vestibularis. **b** Higher magnification of area outlined in **a**. *ar*, anterior division of the eighth nerve; *pr*, posterior division of the eighth nerve; *NA*, nucleus angularis; *NML*, nucleus magnocellularis lateralis; *NMM*, nucleus magnocellularis medialis. From Miller (1975) **a** × 30; **b** × 100

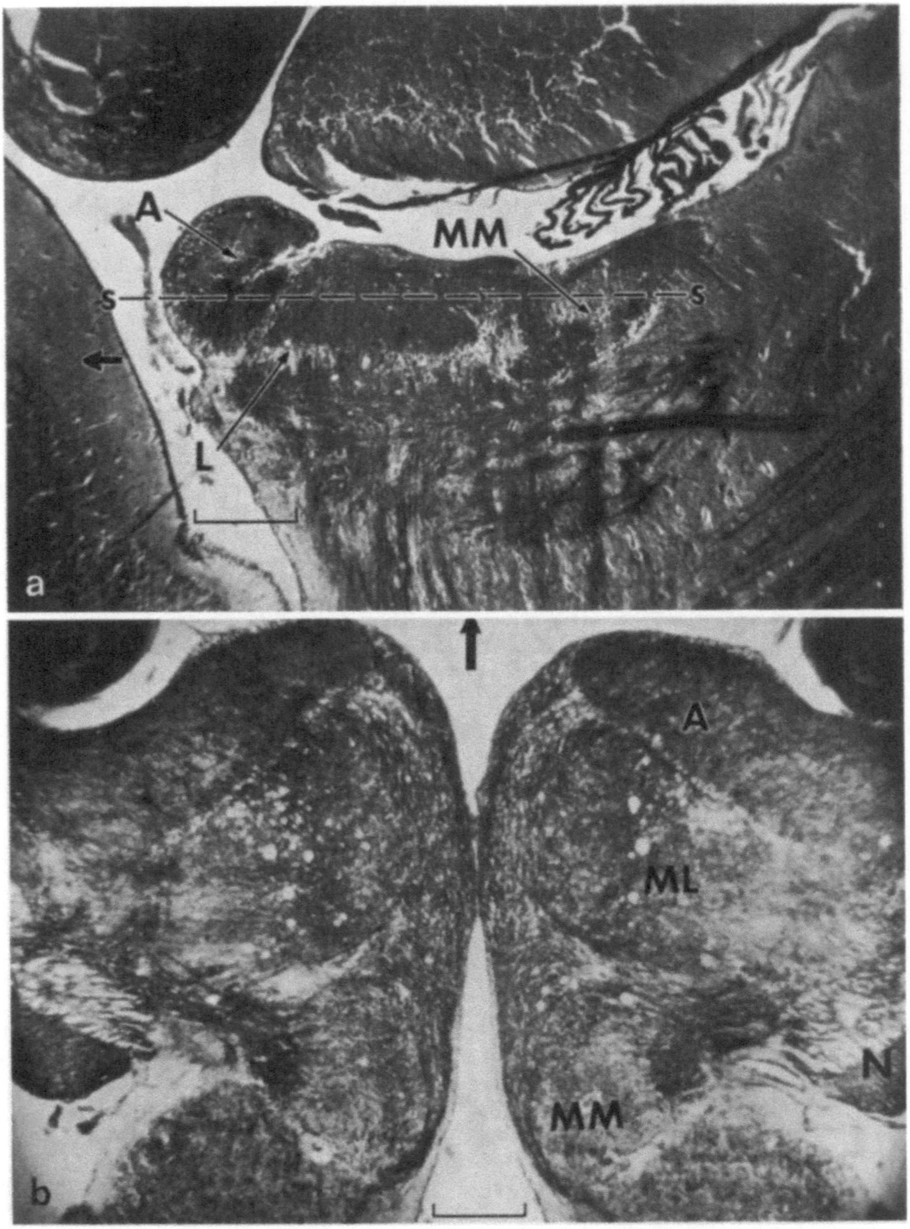

Fig. 43 a, b. Photomicrographs of the cochlear nuclei in the caiman *Caiman crocodilus*. **a** Saggital section through the anterior medulla. Scale line is 500 μ. Arrow points rostrally. *A*, nucleus angularis; *L*, nucleus laminaris; *MM*, nucleus magnocellularis medialis. **b** Frontal (horizontal) section through the cochlear nuclei at the depth indicated by the line s – s in **a**. Scale line is 200 μ. *ML, nucleus magnocellularis lateralis; N*, auditory nerve. Arrow points rostrally. From Manley (1970a)

others) cross connections exist mainly between anterior nuclei on both sides, and this makes homologization impossible. In this group he uses the older terms "nucleus anterior" and "nucleus posterior."

In *Caiman,* Leake (1974) and Manley (1970a) defined the NA as occupying the cephalic fourth of the acoustic tubercle (Fig. 43). It lies anterior to the posterior root of the eighth nerve. The nucleus laminaris is a thin, curved sheet of cells in the anterior half of the tubercle, but probably does not receive primary cochlear fibers, NML overlies NL but extends further laterally. NMM, medial and caudal to NML, has large cells which are arranged in vertical columns in the medial portion of the nucleus, adjacent to the fourth ventricle. Glatt (1975a) reaches similar conclusions for *Caiman,* albeit providing a more complex division of the nuclei. In turtles, Miller and Kasahara (1979) were able to define NA and NMM, but could not distinguish a NML. The observation that NL was to be distinguished in these species led these authors to the suggestion that a well-developed NL is found in those forms with well-developed hair cell unidirectionality, that is, turtles, crocodilians, and birds. In the teiid lizard *Ameiva ameiva,* where there is a large percentage of unidirectionally orientated hair cells, NL is easily definable (Miller and Kasahara 1979).

5.3 Physiology

Suga and Campbell (1967) reported the first recordings from single neurons in the reptile cochlear nucleus (CN) (and possibly superior olive). In the gecko *Coleonyx variegatus,* they found that the cells were sensitive and frequency selective to a high degree and had V-shaped tuning curves, and CFs ranging from 110 Hz to 4.0 kHz (Fig. 44). From these data they were able to construct an audiogram for this species which indicated sensitivity up to about 10 kHz. In ad-

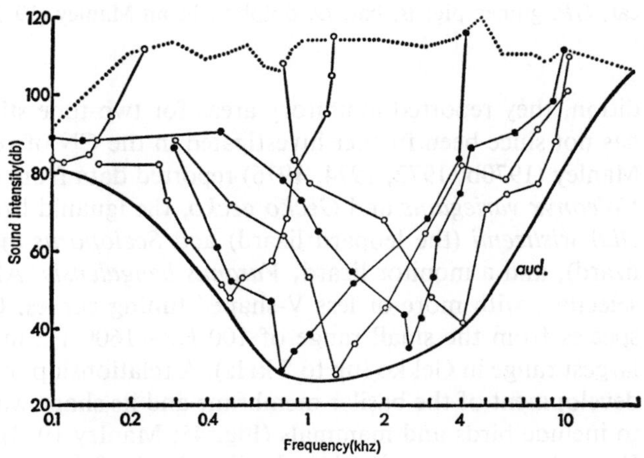

Fig. 44. Tuning curves of six single auditory neurons in the medulla of *Coleonyx variegatus,* the western banded gecko. The curve labeled *aud.* is the audiogram suggested from the tuning curves of all neurons studied and a threshold curve for multiunit evoked potentials. Ordinate dB re 20 μ Nm^{-2}. From Suga and Campbell (1967) Copyright 1967 by the American Association for the Advancement of Science. Used by permission

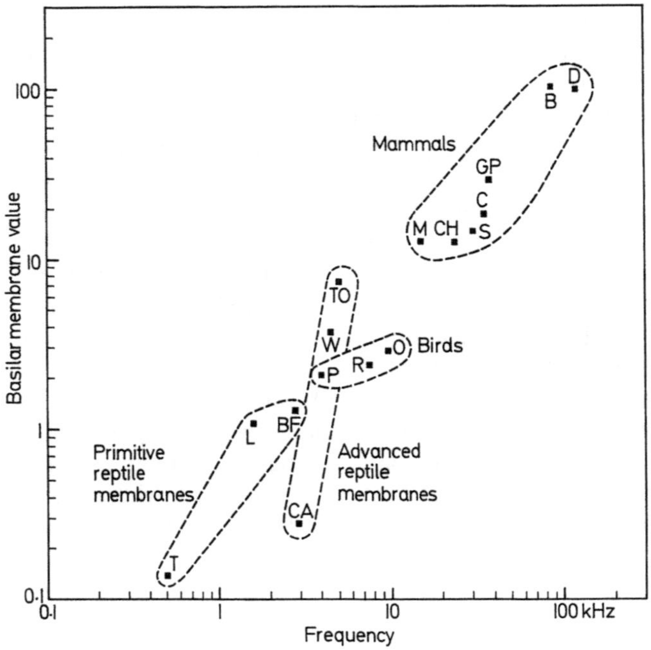

Fig. 45. Correlation between the basilar membrane dimensions and the highest CFs of single neurons in various terrestrial vertebrates. "Basilar membrane value" is the ratio of the membrane shape [length (L) divided by widest width (Ww)] to its width gradient [widest width times narrowest width (Wn)]. The value is calculated as L/Ww: Ww × Wn in millimeters divided for convenience by 100. High membrane values indicate long, thin membranes and vice versa. A simple linear regression analysis yielded a correlation coefficient of 0.973 and a Student's t test value significant at the 0.001 level. *T*, box turtle; *L*, leopard lizard; *BF*, blue fence lizard; *CA*, caiman; *W*, western banded gecko; *TO*, tokay gecko; *P*, pigeon; *R*, American robin; *O*, barn owl; *M*, man; *CH*, chinchilla; *S*, sheep; *C*, cat; *GP*, guinea pig; *B*, bat; *D*, dolphin. From Manley (1973)

dition, they reported inhibitory areas for two-tone stimuli. This latter question has not since been further investigated in the CN of reptiles.

Manley (1970b, 1972, 1974, 1976) reported data from CN neurons in the geckos *Coleonyx variegatus* and *Gecko gecko,* the iguanid lizards *Crotyphytus (Gambelia) wislizenii* (the leopard lizard) and *Sceloporus cyanogenys* (the blue fence lizard), and a monitor lizard, *Varanus bengalensis.* All neurons were frequency selective, with more or less V-shaped tuning curves. CF ranges varied between species from the small range of 100 Hz – 1600 Hz in the leopard lizard to the largest range in Gekko (up to 5 kHz). A relationship of frequency response to the development of the basilar membrane and its shape was calculated and extended to include birds and mammals (Fig. 45; Manley 1973).

From these comparative data for lizards, the following generalizations emerged. Firstly, the tuning curves of some species had no particular asymmetry as, e.g., in *Varanus,* but in other species were mostly asymmetrical (Fig. 46) having a steeper slope on the high frequency side as, e.g., in *Gekko.* These overall statements tend, however, to obscure the differences seen between units of different CF, as

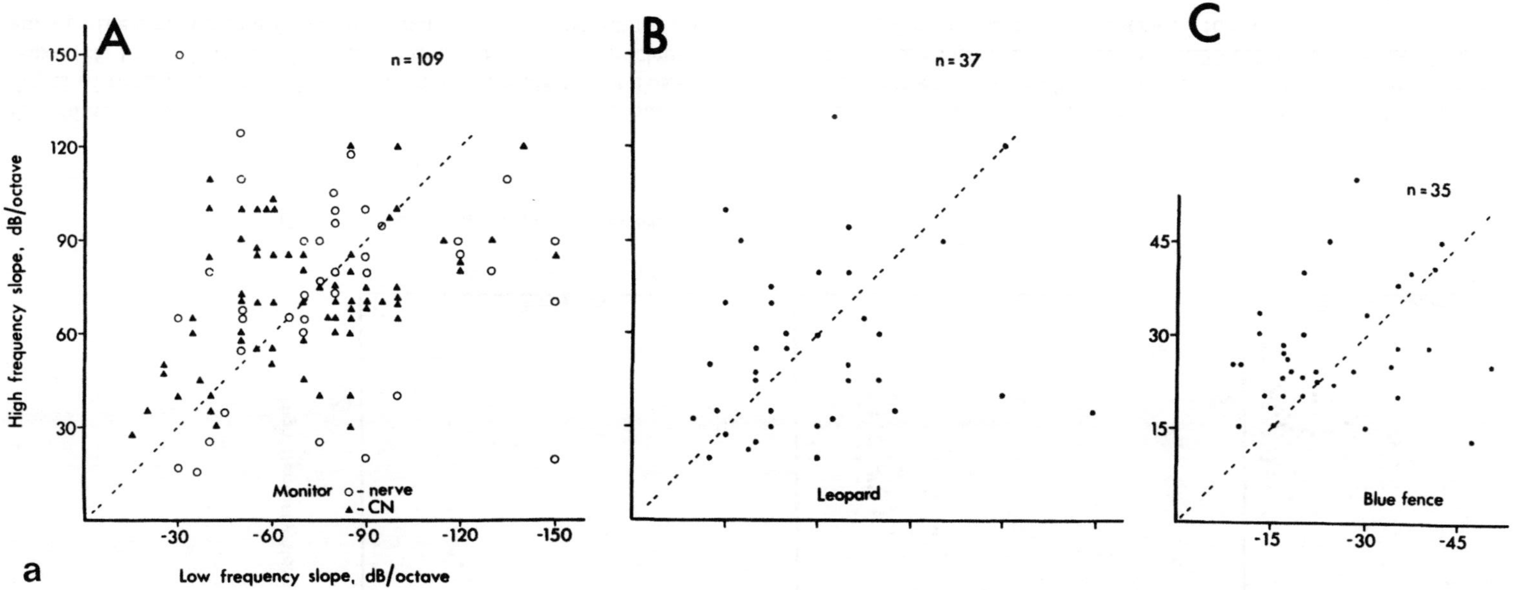

Fig. 46 a. Plots of the high-frequency slopes versus low-frequency slopes of tuning curves of *A*, auditory nerve and cochlear nucleus units in the monitor lizard; *B*, cochlear nucleus units in the leopard lizard; and *C*, cochlear nucleus units in the blue fence lizard. Scales for *B* as in *A*. The *dotted line* in each case indicates symmetrical tuning curves (1:1 slope ratio). The slopes are measured between 3 and 23 dB above CF threshold

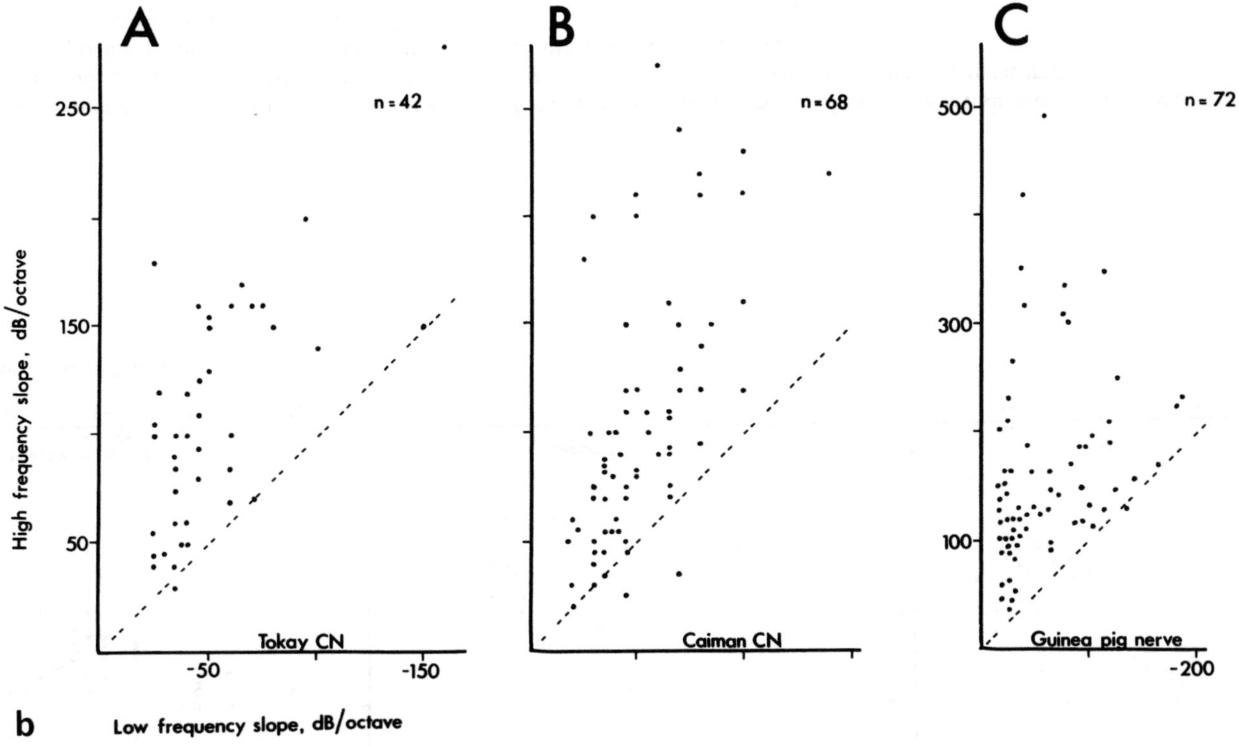

Fig. 46 b. Plots of the high-frequency slopes versus low-frequency slopes of tuning curves of *A*, cochlear nucleus units of the Tokay gecko (cf. Fig. 30b for auditory nerve data); *B*, cochlear nucleus units of the caiman; and *C*, auditory nerve units of the guinea pig (Pawson and Manley, unpublished data). Scales for *B* as in *A*. The *dotted line* in each case indicates symmetrical tuning curves. Slopes are measured between 3 and 23 dB above CF threshold. The guinea pig data (*C*) is included for comparison already pointed out in Sect. 3.4. Secondly, the tuning curves of neurons receiving input from highly developed papillae tended to have an average Q value which increased with increasing CF. In the other species this was not seen. Thirdly, the data indicated a rough tonotopic organization of the CN of *Gekko* (with lower CFs represented dorsally and caudomedially on the CN) and possibly in other species. The small size of the nuclei and desirability of mapping in a single individual made such studies difficult. However, at least in some species the data indicated a probable tonotopic organization of the peripheral organ. Fourthly, neurons in species with poorly developed papillae tended to fall into more than one CF group, a phenomenon which is probably related to grouping at the periphery. In *Varanus,* the peripheral dichotomy in frequency analysis determines the grouping of units by CF in the CN. It has been noted in Sect. 3.4 that peripheral groupings of hair cells and neurons exist in the alligator lizard. In the blue fence lizard, the peripheral anatomy probably strongly resembles that in the related species *Sceloporus magister* and *S. occidentalis* (Miller 1978a), which have a central short-ciliated unidirectional segment of hair cells with two long-ciliated segments, one apical and one basal. The CN data indicated what appear to be three CF groupings, centred near 800 Hz, 1.8 kHz and 2.5 kHz. It may be that the latter two groupings have their origin in the two separated long-ciliated bidirectional segments. Finally, not all units demonstrated a "primary-like" discharge pattern in response to tone bursts. In many cases good synchrony of the latencies of the first few spikes produced PSTH's with prominent peaks at the beginning. This phenomenon was especially pronounced in *Gekko.* In *Varanus,* the low CF neurons showed primary-like patterns tending at higher CFs to exclusively peaked PSTH's (Fig. 28). The suggestion that this pattern was related to the presence of tectorial sallets in the gecko (Manley 1974) needs to be modified as the tectorial membrane is different in *Varanus.* It is, however, true that the phenomenon is more highly developed in *Gekko.* In *Mabuya* (see below), where a large proportion of the cells are also connected to salletal structures (Fig. 9 shows some skink tectorial structures) high CF cells in the CN were observed to have often a pure "onset" PSTH (see below). Related to the variety in PSTH pattern is an equal variety of input – output or intensity functions (e.g., Fig. 28) where primary-like neurons have steeper intensity functions which reach high discharge levels and "peaked" PSTH units have less steep intensity functions with lower average discharge levels. The slopes of intensity functions tended to be steeper for frequencies below CF and less steep for frequencies above CF, as seen in *Caiman* (Fig. 49). In *Gekko,* some units tested with clicks were able to follow rapid click trains, in some cases responding to 70% of clicks separated by only 1 ms. This ability needs to be studied in nerve fibers of different CF and in different species.

In a single specimen of *Mabuya multicarinata*, a skink, Eatock and Manley (unpublished data) found CFs up to 6.3 kHz, the highest so far reported in lizards. This is not entirely unexpected, as skinks possess highly developed auditory papillae. Below CF 2.5 kHz, these CN tuning curves were a simple V-shape, whereas those above CF 2.5 kHz were less sensitive, more complex in shape and had higher Q values (Fig. 47). Up to CF 1.5 kHz, Q values were between 1.5 and 2.7, whereas above this Q values lay between 3.9 and 13. High-frequency slopes of the tuning curves were in most cases steeper than the low-frequency slopes.

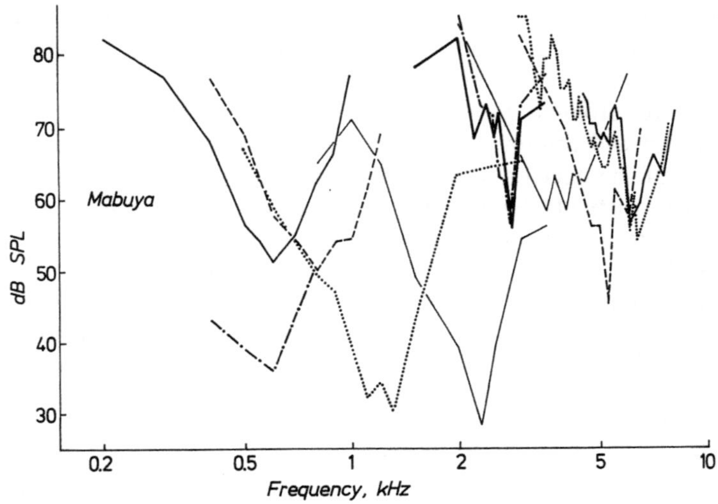

Fig. 47. Tuning curves for units from the cochlear nuclei of the skink *Mabuya multicarinata*

The PSTH shapes for CF tones 20 dB above threshold had broad onset peaks with a lower subsequent discharge for low CF units (0.5 – 0.9 kHz) and showed only an onset pattern for higher CF units (0.8 – 6.5 kHz), similar to a PSTH shown for the nerve of the skink *Trachysaurus* (Johnstone and Johnstone 1969). As some units changed PSTH pattern with intensity, some intensity functions were nonmonotonic. One unit with CF near 5 kHz had an on-off PSTH pattern at frequencies above 6 kHz.

In a brief study of neurons in the CN of the teiid *Tupinambis nigropunctatus*, Browner and Caspary (1976) studied 35 units physiologically. A majority showed primary-like PSTHs, but as the CF range was only 400 Hz to 1.3 kHz, it is likely that a large area of the CN was not examined. From observations on the papilla of the related *T. teguexin* (Miller 1973b) one would expect much higher CFs from a papilla which is probably well developed and predominantly bidirectional.

The above data can now briefly be compared to eighth nerve recordings in two of the species, *Gekko gecko* and *Varanus bengalensis* (see Sect. 3.4). In fact there is little to comment upon in this regard, save to point out that no important differences of any kind have been observed in the two sets of data, although differences would probably emerge if more complex stimuli and other anesthetics were used. This seems to indicate that sound analysis in lizards has a large peripheral component and that the cochlear nuclei probably function mainly as relay stations.

Outside the lizards, data are also available for CN units in a turtle and a caiman (Manley 1970a, b). In the common box turtle, *Terrapene carolina*, the highest CF obtained was 500 Hz, and the units were not very sensitive (best 34 dB SPL). Units were most often encountered and had higher Q values at or near 140 Hz and 400 Hz. That such a grouping is absent in the nerve data from the turtle *Pseudemys scripta* (Paton et al. 1976) may, however, indicate this finding to be a sampling artifact. On the high frequency side, the tuning curves from the CN

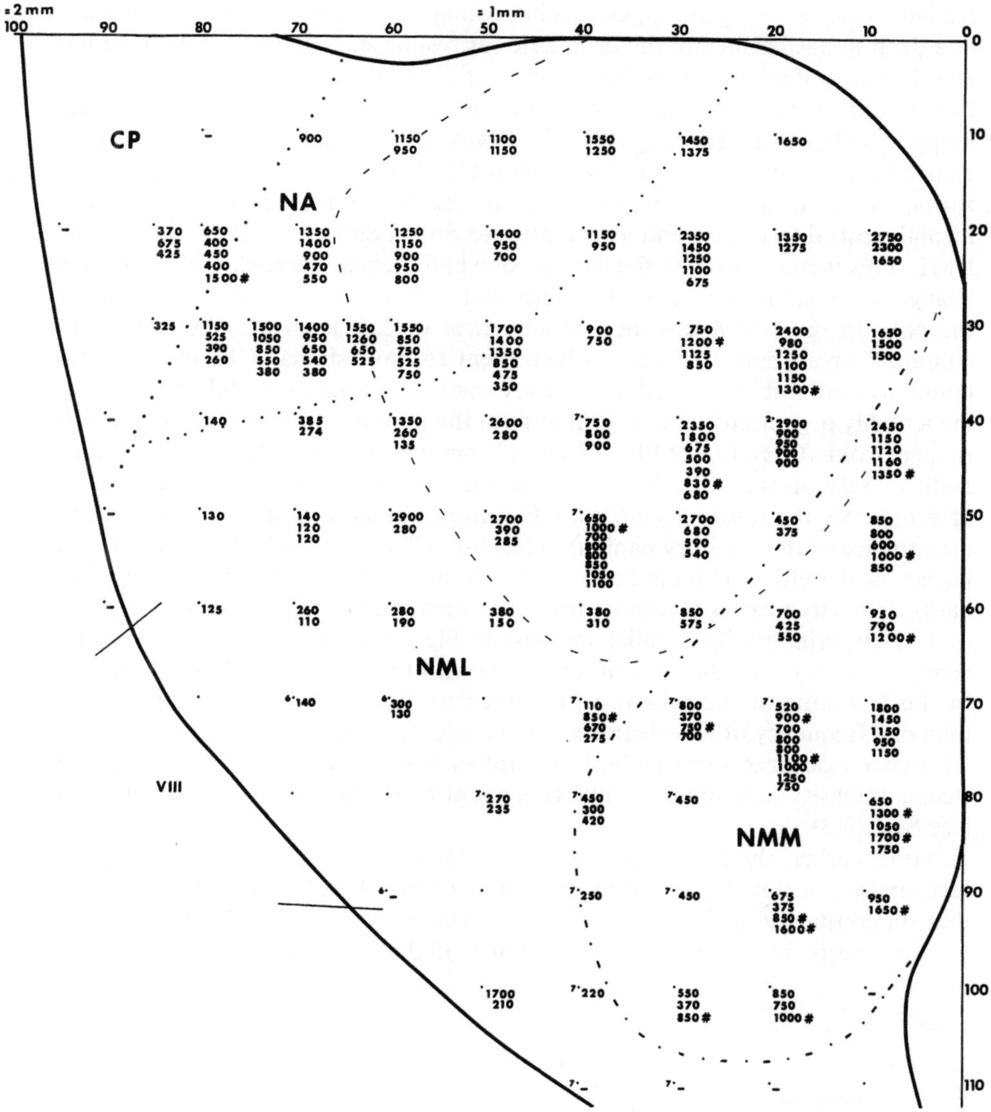

Fig. 48. The tonotopic organization of the left CN of *Caiman crocodilus*. The *dark line* on the right (medially) and anteriorly is the edge of the acoustic tubercle, on the left is the cranial bone. The *dotted line* outlines the NA (nucleus angularis) the dashed line the NMM (nucleus magnocellularis medialis). VIII is at the position of the eighth nerve. Each column of numbers next to the points of electrode penetration shows the vertical sequence of CFs of units encountered. A dash next to the point indicates no auditory units were encountered. From Manley (1970a)

tended to converge together, perhaps indicating a common loss of sensitivity due to a drop in basilar membrane or middle ear response. A similar phenomenon is seen in hair cell and primary fiber responses (Fig. 9b, c).

In the caiman, the enormous size of the cochlear nuclei allowed relatively easy mapping (Manley 1970a) (Fig. 48). The CN is very clearly and systematically organized in a way which in all but one axis perfectly parallels the tonotopic organization of the bird CN (Konishi 1970). In the NA, CFs increase in lateral to medial, rostral to caudal and also ventral to dorsal directions. In the NMM and NML, CFs increase toward the rostral edge and medial margins, with no abrupt change between the two nuclei. No consistent vertical organization is seen outside the NA. In the NMM this lack of a vertical organization may be due to the columnar arrangement of cells, which seem to have similar CF. In birds, the lateral to medial shift upward of CF is reversed. These data were taken as indicating a highly organized frequency analysis at the periphery. CFs in the caiman CN ranged from 70 Hz to 2.9 kHz. Bilateral removal of the cochleas abolished all auditory responses in the CN, while removal of only the lagenar otolith material affected only the *sensitivity* of low CF auditory units (presumably due to some disturbance of the auditory papilla), while not affecting the CF distribution. The lagena is therefore unlikely to be auditory in function (Manley, unpublished data). Discharge patterns to pure tones in *Caiman* CN units (Manley 1974) were exclusively primary-like (similar to those in Fig. 41A), with high maximum discharge rates (to over 200/s). This contrasts with the PSTH variability in lizards.

In the few units of the caiman where intensity functions were studied at more than one frequency, it was clear that frequencies lower than the CF often produced higher discharge rates at high intensities than at the CF itself, and had a steeper intensity function (Fig. 49). This has also been noted for the caiman nerve (see Sect. 3.3).

As noted earlier, the development of the NMM is related to the size of the unidirectionally orientated hair cell areas (Miller 1978a). It is thus of interest to note that, in conjunction with the physiological evidence that the unidirectional hair cell areas respond to low frequencies, almost all the high frequency units in *Vara-*

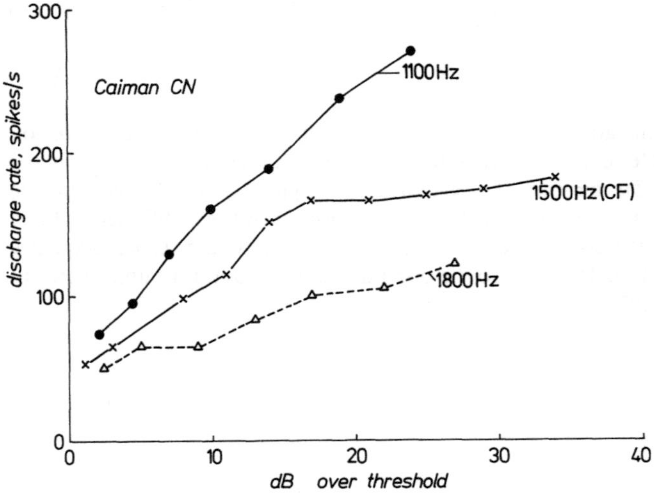

Fig. 49. Intensity functions for a cochlear nucleus unit in *Caiman crocodilus,* at three frequencies: CF (1500 Hz), below CF at 1100 Hz and above at 1800 Hz. "dB over threshold" refers to the threshold for each frequency so that slopes are more easily compared

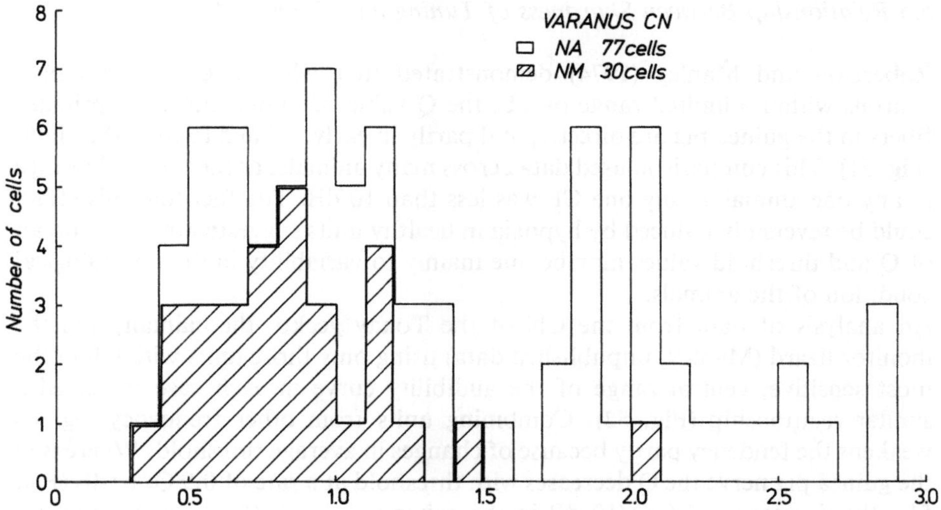

Fig. 50. Number of cells in the cochlear nuclei of *Varanus* with CFs in different ranges (100-Hz divisions of the range). The ***hatched area*** represents cells which are localized during the experiment as falling outside the NA swelling. Virtually no high-CF units were found in the NM

nus CN (Manley 1976) were identified at the time of recording as coming from the NA (Fig. 50). Similar results are available for the CN of the Tokay gecko. Such a finding was not obvious in data from the caiman (Manley 1970a), where the highest frequencies originate in NML. However, in birds (Konishi 1970) the very highest frequencies (above 5 kHz) are only found in the most lateral area of NA. It is difficult to estimate to what extent these frequency distribution patterns in the CN are the result of common innervation patterns. The fact that the frequency distribution on the basilar papilla in *Varanus* is complex (Manley 1976), and is in *Gekko* probably the reverse of mammals and birds, complicates the matter.

5.4 Spontaneous Activity in the Cochlear Nuclei

Attempts to define the overall rate distribution of spontaneous activity in the CN have been less than fruitful due, firstly, to the fact that the presence of a pressure field from the electrode directly affects the discharge rate (Manley 1972c). Secondly, in our experience not only the dosage but also the rate of application of the dosage of the Urethane (ethyl carbamate) anesthetic affects the overall activity of the eighth nerve and CN. It is of more interest to consider the spontaneous activity patterns of the primary fibers (see Sect. 3.3), although here also an effect of anesthesia is probable.

5.5 Relationship Between Sharpness of Tuning and Threshold

Robertson and Manley (1974) demonstrated that when one considers only neurons within a limited range of CF, the Q value of tuning curves of primary fibers in the guinea pig are directly, and partly linearly, related to their threshold (Fig. 51). This comparison used data across many animals, as the threshold range in any one animal at any one CF was less than 10 dB. The fact that this effect could be reversibly induced by hypoxia in healthy units suggested that the spread of Q and threshold values may be due mainly to variability in the physiological condition of the animals.

An analysis of data from the CN of the Tokay gecko, the caiman, and the monitor lizard (Manley, unpublished data) using only those units with CF in the most sensitive, central range of the audibility curve in each case indicated a similar relationship (Fig. 52). Combining units from other frequency regions weakens the tendency partly because of changes in average threshold. Whereas in the guinea pig nerve the Q decreases with threshold at a rate of 0.9 Q/10 dB (Fig. 51), the decrease is 1.0 Q/10 dB in the caiman, 1.11 Q/10 dB in the monitor lizard, and 0.98 Q/10 dB in the Tokay gecko (Fig. 52). In the gecko nerve, all units together showed a weak tendency in the same direction (Eatock 1978). These data possibly indicate an additional respect in which the frequency tuning properties of reptile and mammal auditory systems resemble each other.

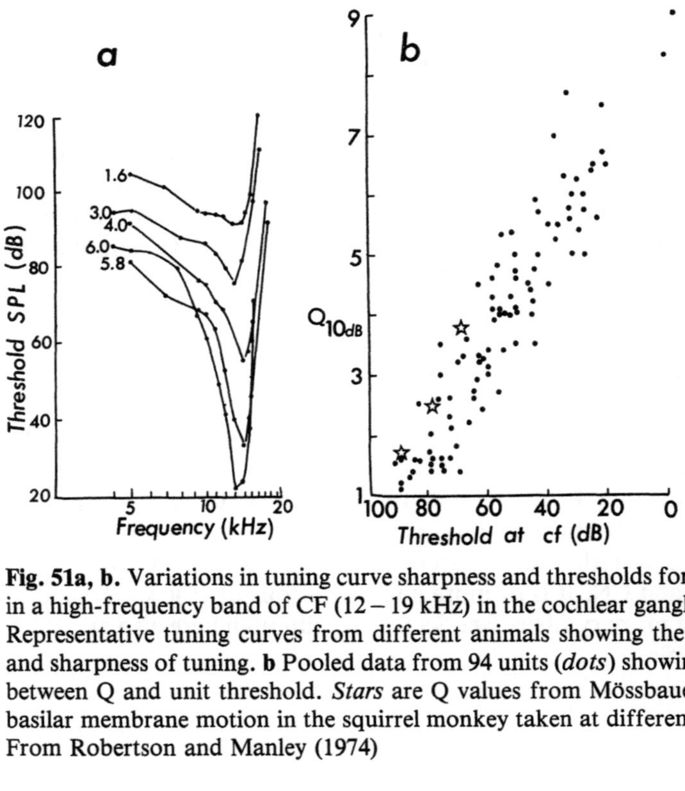

Fig. 51a, b. Variations in tuning curve sharpness and thresholds for primary auditory units in a high-frequency band of CF (12 – 19 kHz) in the cochlear ganglion of the guinea pig. **a** Representative tuning curves from different animals showing the gradation in threshold and sharpness of tuning. **b** Pooled data from 94 units (*dots*) showing the close relationship between Q and unit threshold. *Stars* are Q values from Mössbauer measurements of the basilar membrane motion in the squirrel monkey taken at different sound pressure levels. From Robertson and Manley (1974)

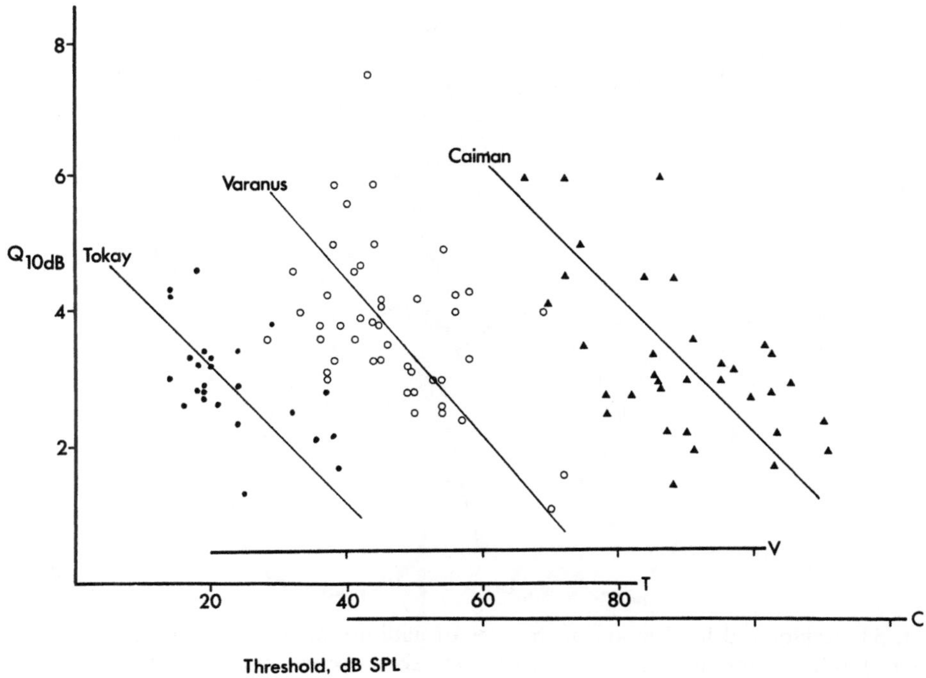

Fig. 52. Variations in tuning curve sharpness and threshold for cochlear nucleus units in the Tokay gecko (*abscissa T*; *filled circles*; CFs 1 kHz – 3 kHz); monitor lizard, *Varanus* (*abscissa V* values displaced 20 dB to the right from *abscissa T*; *open circles*; CFs 600 Hz – 1 kHz); and the caiman (*abscissa C*, a further 20 dB displacement to the right; *filled triangles*; CFs 800 Hz – 1500 Hz). Lines are simple regression fits to the data

6 Higher Centers

6.1 Midbrain

The midbrain auditory nucleus of reptiles is known as the torus semicircularis (Ariens-Kappers et al. 1960; Kennedy 1974; Manley J, 1971; Pritz 1974 a;b). It is divided into an external nucleus continuous with the deep layers of the optic tectum, and a central nucleus. This localization of function compares well with similar studies in fish (Page 1970), amphibians (Potter 1965), and birds (Potash 1970). The relationship between the cochlear nuclei and higher brain centers has been illustrated in Fig. 41 (Foster and Hall 1978).

Kennedy (1974) probed the midbrain area of Tokay geckos with tungsten micro-electrodes and localized multiple-unit responses to white noise pulses within the central nucleus of the torus semicircularis and its boundaries (Fig. 53). Later, Kennedy (1975) reported eliciting vocalizations by electrical stimulation of sites in the peripheral nuclear group of the Tokay midbrain. Similar apposition of auditory and vocalization centers is seen in other vertebrates.

Sammaritano-Klein and Manley (1976 and unpublished data) extended Kennedy's studies by means of single-neuron recordings from the Tokay

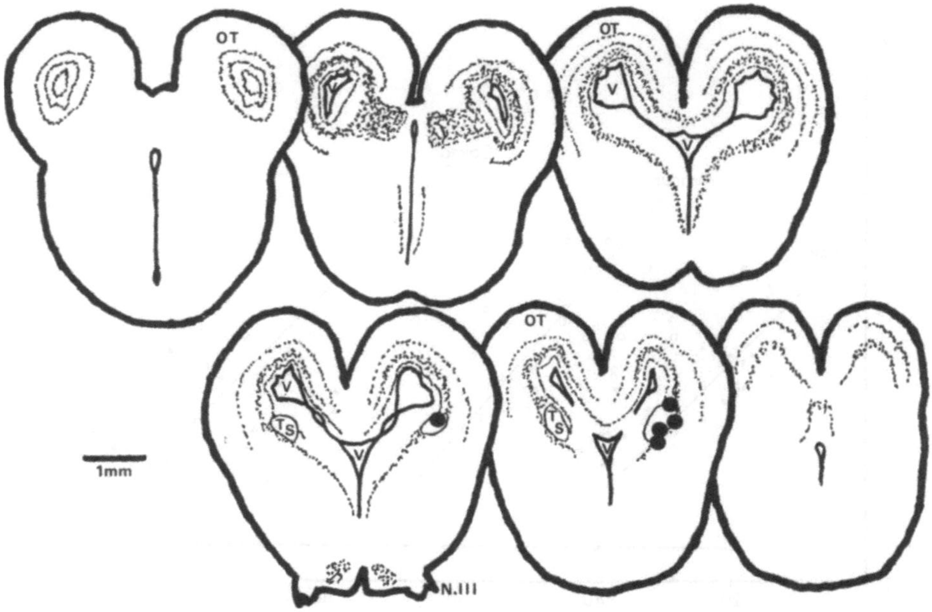

Fig. 53. Anatomical localization of the sites of auditory multiple-unit responses (*filled circles*) in the midbrain of the Tokay gecko. The six transverse section tracings represent levels from rostral (*top left*) through the midbrain at about 400 μm intervals. N.III oculomotor nerve; *OT*, optic tectum; *TS*, torus semicircularis; *V*, ventricle. From Kennedy (1974)

auditory midbrain. Discharge patterns were studied in 87 cells, all localized in the torus semicircularis (Fig. 54). This nucleus was, however, found to be larger than described by Kennedy (1974), who gives its rostrocaudal extent as 800 – 1000 μ. Apparently, Kennedy did not section right to the end of the midbrain (Kennedy, personal communication with Sammaritano-Klein). With coronal and saggital sections, a further posterior portion of the nucleus was describable (Sammaritano-Klein 1976). The bilateral anterior extensions of the nucleus join posteriorly at the midline forming a bilobed nucleus which extends a further 900 μ caudally (Fig. 54).

The range of single-unit CFs in this study was 100 Hz to 4.7 kHz, similar to the range in lower centers. The fact that only 10 % of the units had CFs above 2 kHz can probably to traced to sampling error in a tonotopically organized nucleus. In 8% of cells, adaptation to successive tone bursts occurred and three cells responded only to white noise and not to the tones which were used. Several other cells did not respond at high sound intensities (above about 60 dB). The shapes of tuning curves varied from the simple V-shape characteristic of lower centers to complex shapes like those found in higher centers of other species (e.g., Guinan et al. 1972). Both sharply and broadly tuned units were found (Fig. 55a), the latter often showing secondary increases in sensitivity which in two cells approached in sensitivity that of the CF. Q values showed an increased spread at higher CF, as in lower centers, with values at high CF between 1 and 12, with relatively more broadly tuned units than seen in the nerve or CN. In 18 % of

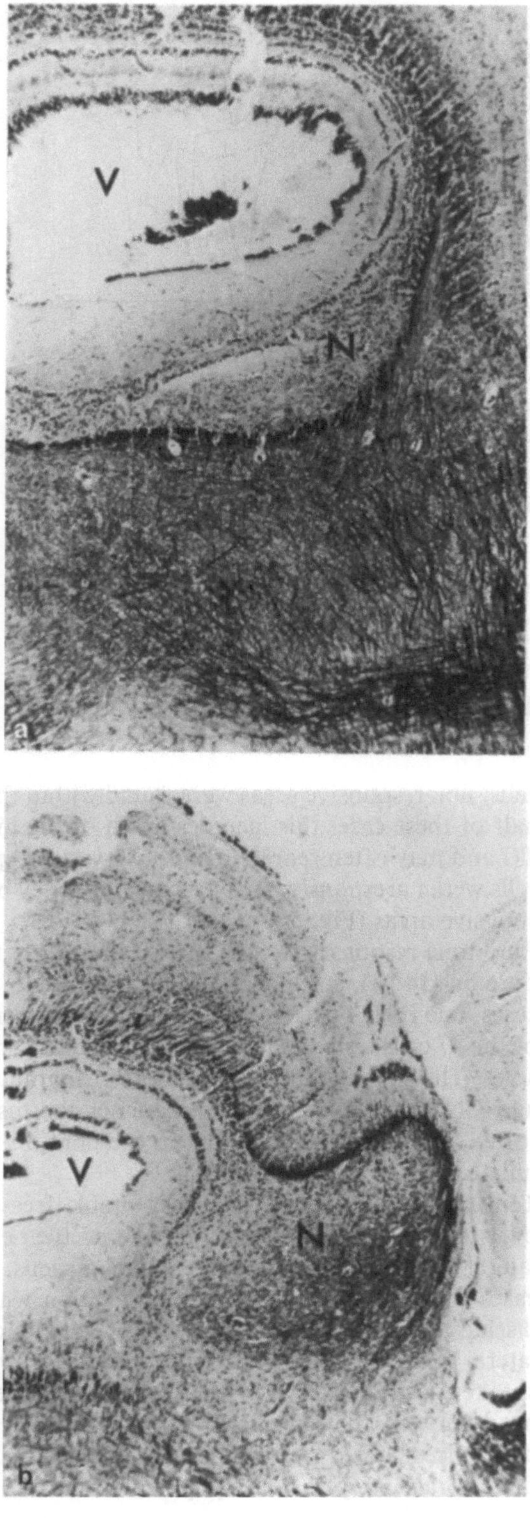

Fig. 54a, b. Photomicrographs of the torus semicircularis of the Tokay gecko. **a** Saggital section through the midbrain. *N*, nucleus centralis; *V*, ventricle. Location of section: 0.5 mm medial from the edge of the midbrain. Dorsal is up. **b** Same, as in **a**, 1.0 mm medial from the edge of the midbrain. Sammaritano-Klein and Manley, unpublished data

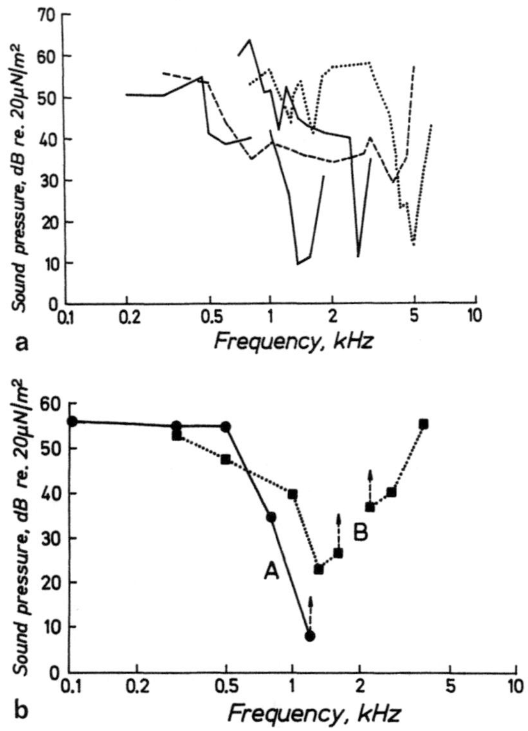

Fig. 55. a Tuning curves from single units in the torus semicircularis of the Tokay gecko midbrain, selected from one individual, to show the range of shapes found. **b** Tuning curves from two single midbrain units in the Tokay gecko, with nonresponsive areas. The high-frequency slope in unit **a** was not definable. Unit **b** was unresponsive near 2 kHz. From Sammaritano-Klein and Manley, unpublished data

cells, non responsive areas were found within the tuning curves, but in more than half of these cases this includes much of the high- or low-frequency part of the TC and may often represent simply a very steep slope of the TC. The remaining cells were unresponsive to a narrow frequency band bounded on both sides by responsive areas (Fig. 55b).

Pure-tone responses were classified according to their PSTH shape: on (17 %), pure on (14 %), sustained (60 %), and late (8 %) responders (Fig. 56). In most cases, one PSTH pattern prevailed at all frequencies and intensities used, but in 7 of the 47 cells tested in this way it shifted to a more onset pattern at higher intensities. The sharp onset peak of most histograms resembles that seen in the nerve and CN. Increasing intensity increases the discharge rate over a certain range (often less than 10 dB), except in the pure on-responders, which normally gave only one spike per presentation.

Ten units were studied using a rotatable free-field loudspeaker and presenting pure tones at different positions around the head, to favor in turn putative ipsilateral and contralateral inputs to this nucleus. In these units, the PSTH in most cases changed as the loudspeaker location was changed, such that the highest discharge activity was usually seen with the loudspeaker at a position 90° contralateral to the nucleus, and latency increased several milliseconds at the ipsilateral speaker position (Fig. 57). Measurements of the acoustic shadowing effect of the head showed that over the frequency range of interest to these species, no differences between the ears exceeded 8 dB, and values of 2 – 3 dB, i.e., near the limit of accuracy of the equipment, were more normal. These small intensity

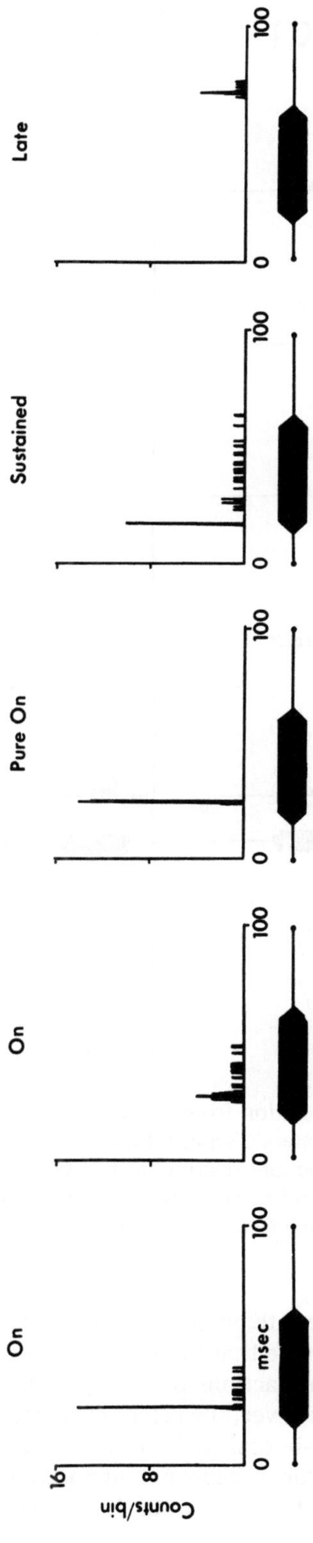

Fig. 56. PSTHs for CF stimulation, 20 dB above threshold, repeated 50 times for five units in the midbrain of the Tokay gecko to illustrate the four PSTH categories. From Sammaritano-Klein and Manley, unpublished data

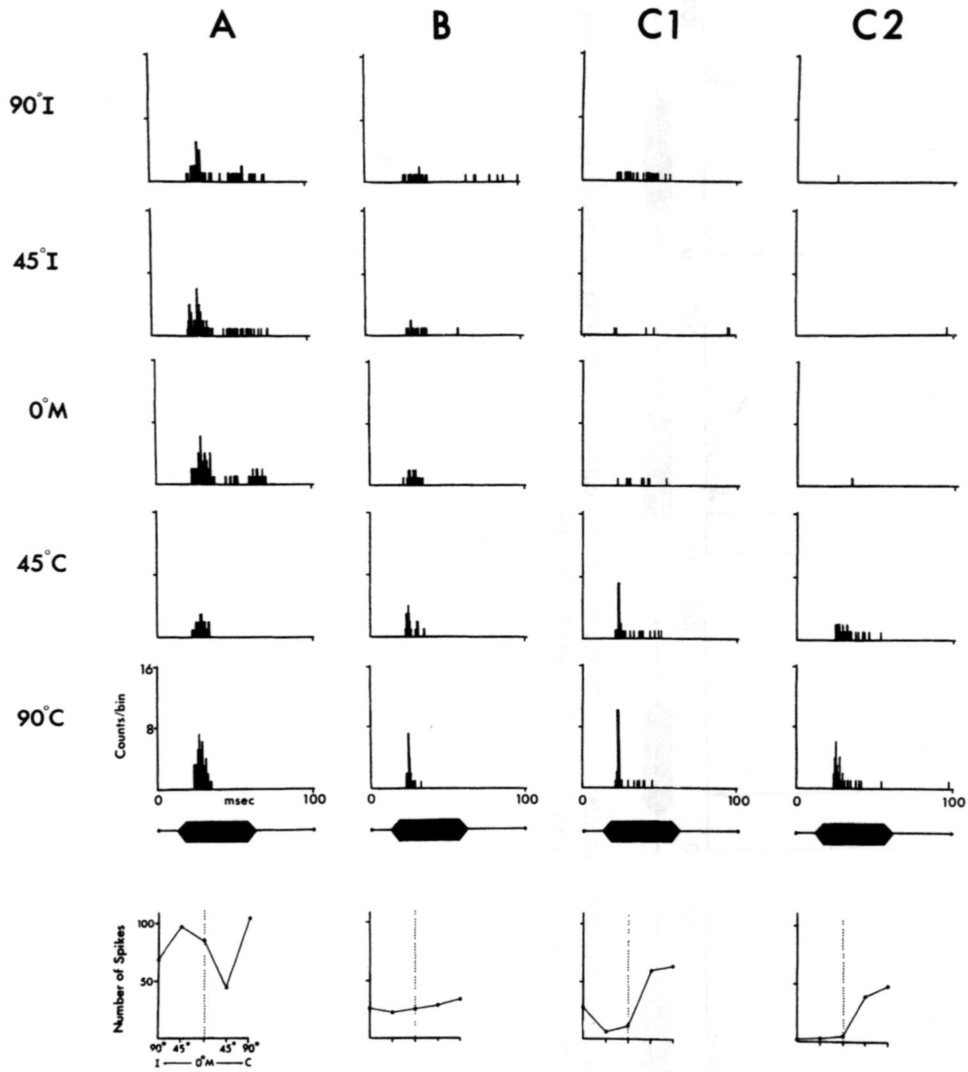

Fig. 57. PSTHs for three units *A, B,* and *C* at CF, for stimulation from the loudspeaker positions 90° contralateral through 45° intervals to 90° ipsilateral. Series *C1* and *C2* were taken at 30 and 20 dB above threshold, respectively. The functions relating the number of spikes (to 50 tone-burst presentations) to loudspeaker position for each case are shown below the histograms, scales at *left*. From Sammaritano-Klein and Manley, unpublished data

changes cannot produce the changes seen in discharge rate based on predictions from intensity functions, so it seems reasonable to assume that these cells are binaurally innervated, with excitatory and inhibitory interactions occurring. This was not further studied using two speaker systems. However, the recent evidence for interaural effects through the common middle ear-buccal cavity (see Sect. 2) make it seem likely that some of these effects are not due to neuronal interaction but to middle ear interference. The fact that most cells behaved similarly, in con-

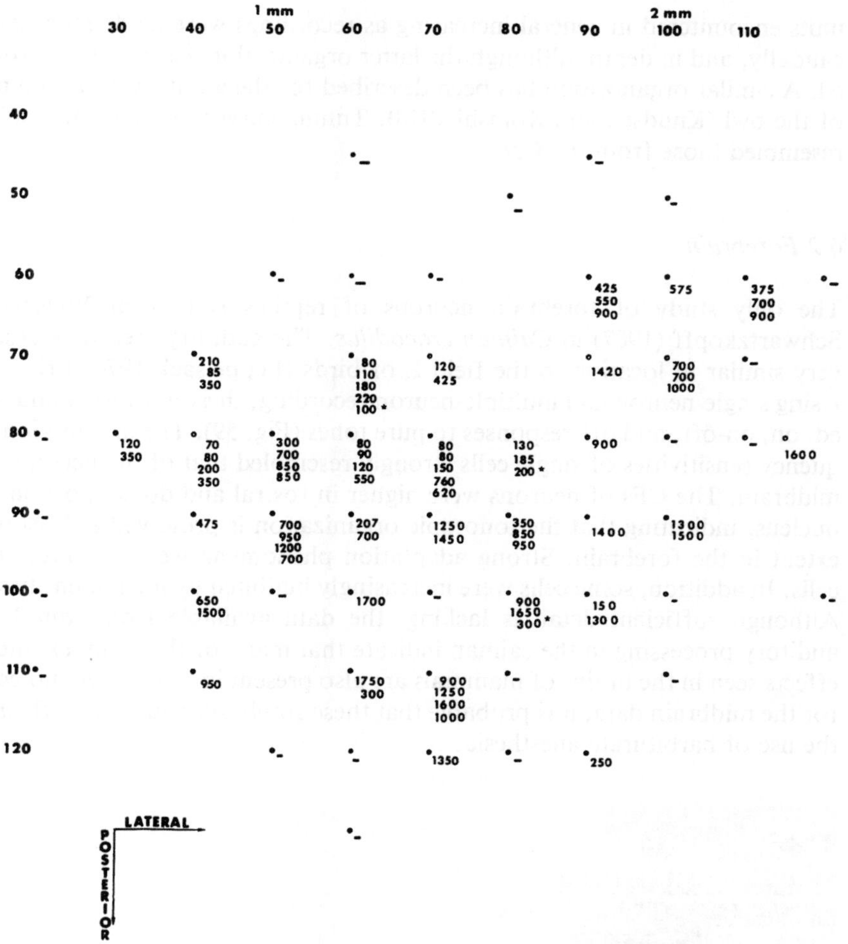

Fig. 58. The tonotopic organization in the right auditory nucleus of the caiman midbrain. The *coordinates* represent distances laterally from the midline and posteriorly from the junction of fore- and midbrains. Each column of numbers represents the vertical sequence in which units of those CFs were found at locations indicated by the *dots*. A *dash* indicates that no auditory units were encountered at that location. An *asterisk* indicates a sustained unit. From J. Manley (1971)

trast to the mixed situation in the cat midbrain nucleus (Guinan et al. 1972), speaks perhaps for a relatively strong middle ear effect.

In the midbrain of the caiman, J. Manley (1971) reported data from 206 single auditory units in the torus semicircularis. Of these units, about 90 % gave phasic responses, only firing at the onset of the stimulus. Apart from one "off" firing unit, the rest gave sustained responses. The range of CFs and sensitivities was similar to that encountered in the CN of this species (Manley G 1970a). Recent data from bird and mammal higher centers (e.g., Sachs and Sinnott 1978) indicate that the barbiturate anesthesia has a strong influence on the response patterns seen. The caiman torus was also tonotopically organized (Fig. 58), with the CF of units encountered in general increasing as recordings were made more laterally, caudally, and in depth, although the latter organization was the most pronounced. A similar organization has been described for the auditory midbrain nucleus of the owl (Knudsen and Konishi 1978). Tuning curves in the caiman midbrain resembled those from the CN.

6.2 Forebrain

The only study of forebrain neurons of reptiles is that of Weisbach and Schwartzkopff (1967) in *Caiman crocodilus*. The auditory area they describe is very similar in location to the field L of birds (Leppelsack 1974; Pritz 1974b). Using single-neuron and multiple-neuron recording, these authors found sustained, on, on-off, and off-responses to pure tones (Fig. 59). The distribution of frequency sensitivities of single cells strongly resembled that of the nerve, CN, and midbrain. The CFs of neurons were higher in rostral and dorsal portions of the nucleus, indicating that the tonotopic organization is preserved at least to some extent in the forebrain. Strong adaptation phenomena were observed in some cells. In addition, some cells were increasingly inhibited by high intensity sounds. Although sufficient detail is lacking, the data available from four levels of auditory processing in the caiman indicate that many of the complex interactive effects seen in the brains of mammals are also present in reptiles. As noted above for the midbrain data, it is probable that these forebrain data were influenced by the use of barbiturate anesthesia.

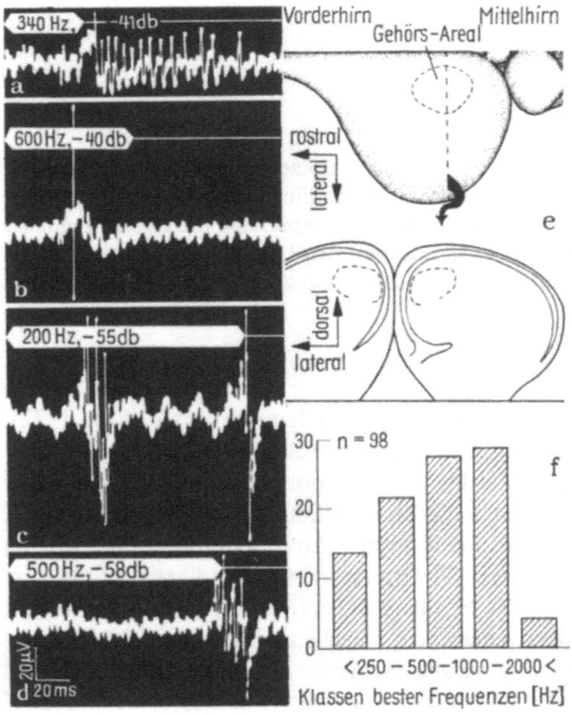

Fig. 59 a – f. Representative activity patterns, location, and best frequency distribution of recordings from an auditory area in the forebrain of the caiman. **a** Sustained, rhythmic discharge. **b** Early response. **c** Early and late response. **d** Late response only. **e** Auditory area outlined in the forebrain seen from above and in transverse section. **f** Best frequencies of auditory neurons classified into five frequency groupings. From Weisbach and Schwartzkopff (1967)

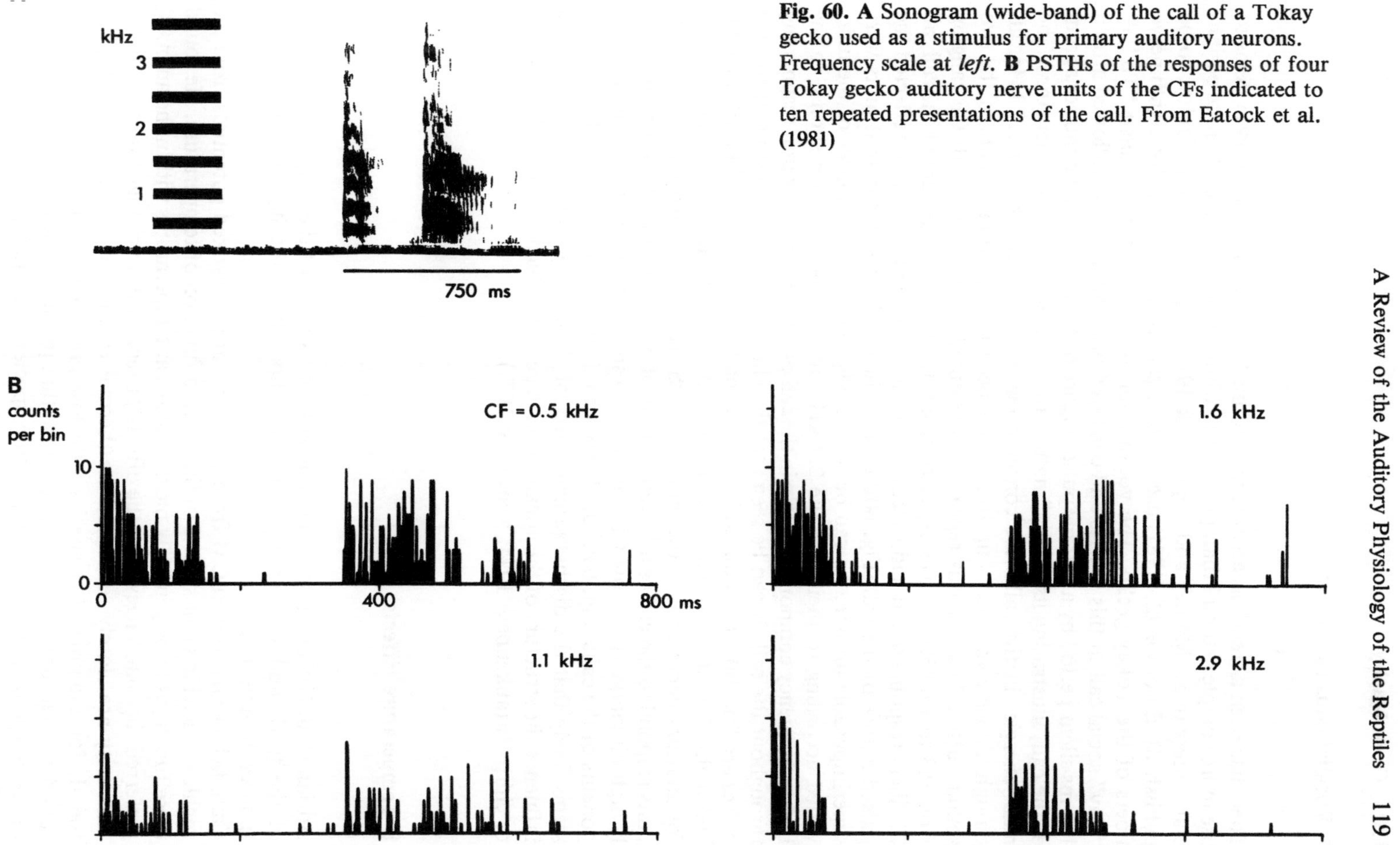

Fig. 60. A Sonogram (wide-band) of the call of a Tokay gecko used as a stimulus for primary auditory neurons. Frequency scale at *left*. **B** PSTHs of the responses of four Tokay gecko auditory nerve units of the CFs indicated to ten repeated presentations of the call. From Eatock et al. (1981)

7 Vocalizations as Stimuli

There has been a trend in recent studies of bird and mammal auditory systems to use more complex stimuli than pure tones, white noise, etc., in even quite peripheral centers (e.g., Manley and Leppelsack 1977, Leppelsack and Manley, to be published). Eatock et al. (1981) used a vocalization as a stimulus for eighth nerve fibers of the Tokay gecko. The vocalization was the two-component "To" – "Kay" social call of this species and consisted of a 150-ms pulse followed after a 190-ms silent period by a 400-ms pulse. Figure 60A shows a wide-band sonagram of the call, illustrating its "noisy" nature and its pronounced time structure. The main energies in the call were concentrated at frequencies below 1.7 kHz, although higher components up to 3.5 kHz occurred in the early part of the two sound pulses. The call was repeatedly presented to 20 fibers, with CF ranging from 0.3 to 3.6 kHz. As was expected from the broadband nature of the call, all 20 fibers responded and had successive peaks in the PSTH's which reflected the pulsed nature of the call (Fig. 60B). Peaks in PSTHs from different units were superimposable in time. As can be seen in Fig. 60 B, the responses of the units varied according to their CF, the 2.9 kHz unit giving a short response to the higher frequency components early in the call pulses, the 0.5 kHz unit responding throughout the call. It can be seen that the click-like subcomponents of the call pulses are faithfully reproduced in the time patterning of the PSTHs of the mid- and high-CF units. This click-like pattern in fact provides an optimal stimulus for auditory-nerve units in the onset-peak PSTH category (see Sect. 3.4). These units respond to the call as if it were a series of onsets and thus average a relatively high discharge rate. Low-CF units tend to smear together the individual components of the calls (Fig. 60B, 0.5 kHz) with their primary-like PSTH pattern. It seems likely that the discharge properties of the primary auditory nerve fibers influence the structure of the vocalization in geckos, which tend to be brief pulses or chirps (Frankenberg 1974; Marcellini 1977).

8 Temperature Effects

Adrian et al. (1938) reported a temperature sensitivity of the frequency response of the turtle auditory nerve. At lower temperatures the higher frequency responses disappeared.

Campbell (1969) studied CM from the inner ear and evoked potentials from the cochlear nuclei in eight species of lizards as a function of temperature. He found these functions to be generally most sensitive at temperatures within the animals' preferred thermal ranges, i.e., the effects of temperature change on the auditory sensitivity are closely related to the thermal ecology of the species. Campbell found, for example, that species with low preferred activity temperatures respond at their maximum at temperatures that produce a profound depression in the response of the higher temperature species. Generally, the midfrequencies (1 – 2 kHz) were most strongly affected by a temperature change.

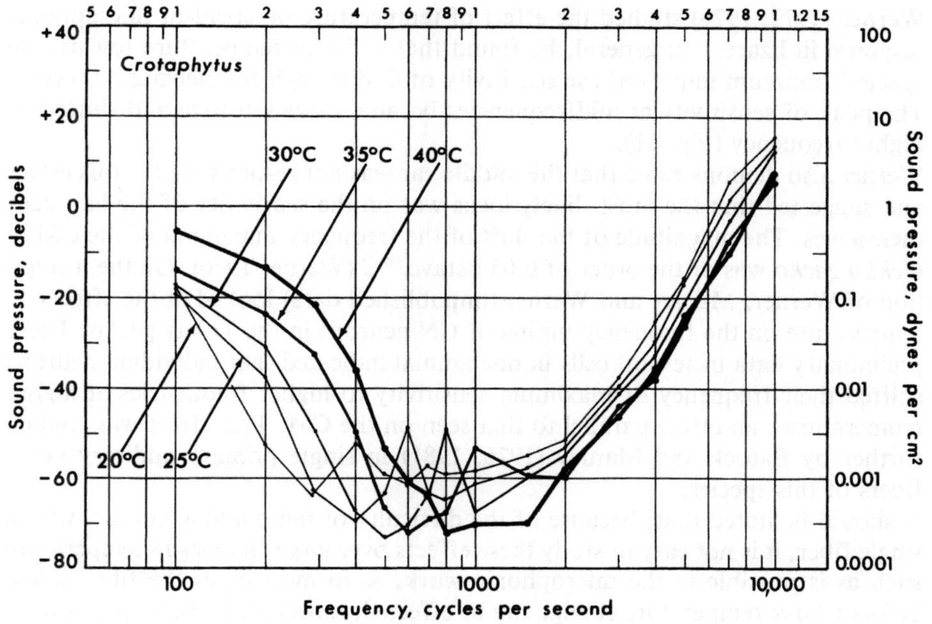

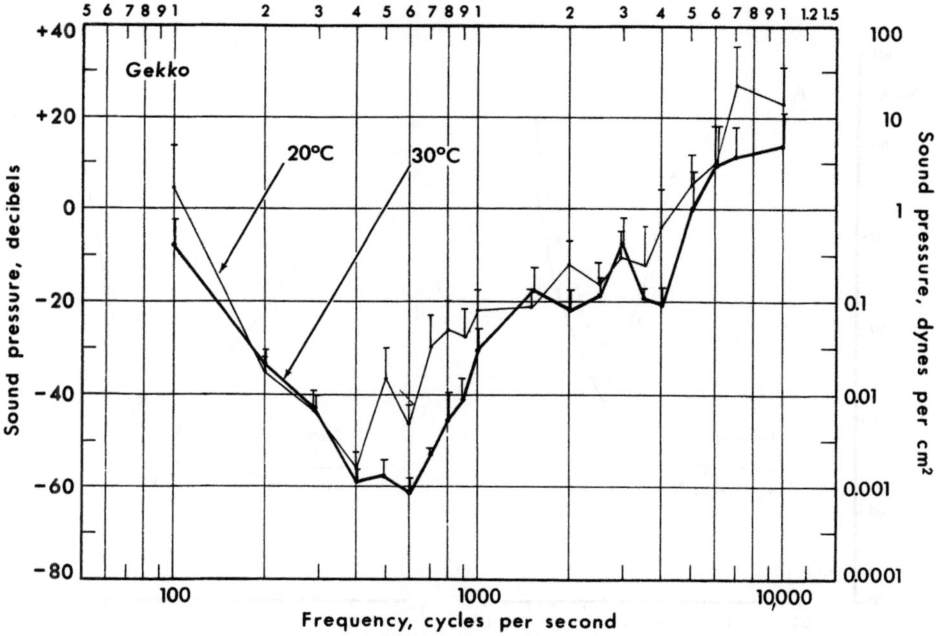

Fig. 61. *Above:* Five CM sensitivity functions for the iguanid lizard *Crotaphytus collaris* obtained at the temperature indicated, all from one specimen, illustrating the shift in the peak frequency with temperature. From Werner (1972). *Below:* Averaged CM sensitivity functions for the Tokay gecko at two temperatures, showing the peak frequency shift with temperature. Also plotted is one standard deviation for each point. From Werner (1976)

Werner (1972, 1976) studied the effect of temperature on cochlear microphonic response in lizards. In general, he found that a rise in temperature toward the species' optimum improved the sensitivity of CM at high frequencies markedly. The peak of sensitivity at midfrequencies became more sensitive and shifted to higher frequency (Fig. 61).

Werner also demonstrated that the middle ear was not responsible for this effect and suggested that the most likely locus was on the sensitivity of the hair cells themselves. The magnitude of the shift of the frequency maximum of the CM in *Gekko gecko* was of the order of 0.05 octave/°C (Werner 1976). On the instigation of Werner, Manley and Werner (unpublished data) looked at the effects of temperature on the frequency tuning of CN neurons in the Tokay gecko. These preliminary data in several cells in one animal indicated that individual neurons shifted their frequency of maximum sensitivity to higher frequencies at higher temperatures, an effect parallel to that seen on the CM. This effect was studied further by Eatock and Manley (1976, 1981) in single primary auditory nerve fibers of this species.

It should be noted that, because of the difficulty of maintaining contact with a single fiber, it is not easy to study these effects over large changes in temperature such as is possible in the microphonic work, as in most cases the fiber is lost before a large temperature change can be effected, especially if the temperature is raised reasonably slowly. Temperature effects were studied for changes as small

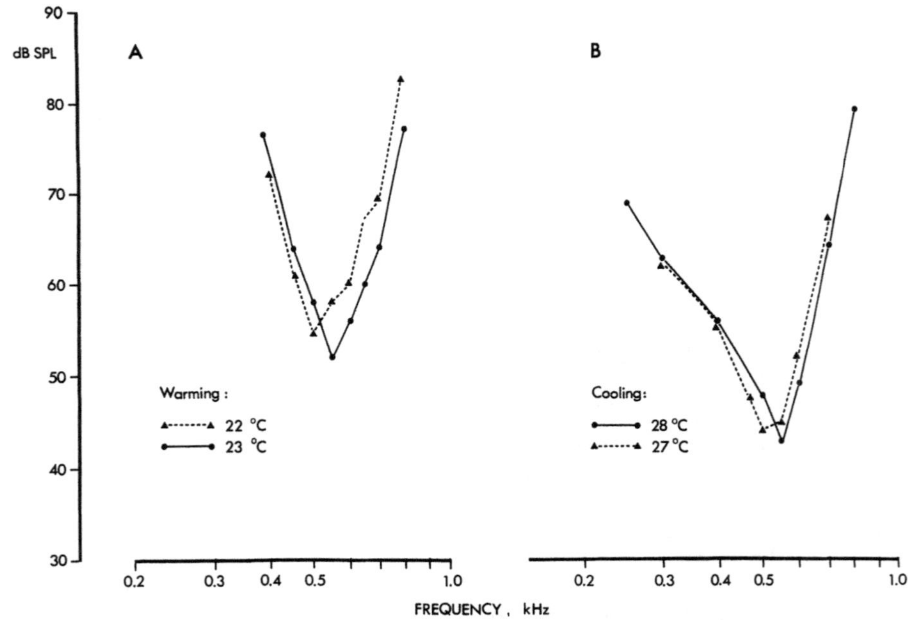

Fig. 62 A, B. Temperature effects on the tuning of two auditory nerve fibers in the Tokay gecko, both having CFs near 500 Hz. **A** Warming the animal from 22° to 23 °C. **B** Curves taken while cooling the animal from 28° to 27 °C (oral temperature). From Eatock and Manley (1981)

as 0.3 °C and as large as 5 °C. The effects shown in Fig. 62 are representative of the results. The CF of the unit in Fig. 62A shifted upward with increasing temperature and the thresholds changed for frequencies other than the CF. The reverse effect is seen for cooling in Fig. 62B. No consistent effect on CF sensitivity was seen, although this may be partly a consequence of the small magnitude or the slowness of the temperature changes studied, or of the limits of the temperature range used (20° – 30 °C). This species has a rather broad temperature tolerance. The only consistent change was a shift of the whole tuning curve as described. In some units held for a long time, this effect was demonstrated to be completely reversible. From the data, it appears as if the actual magnitude (in Hz) of the frequency shift/°C may increase with CF, but it is not possible to state this definitely due to the small sample size at higher CFs. The data are also compatible with a standard shift in frequency of mean value 30 – 40 Hz/°C. This needs to be investigated further with a larger sample of high-frequency units.

Klinke and Smolders (1977), Fengler et al. (1978), and Smolders and Klinke (1977) investigated the influence of head temperature on nerve fiber tuning in the caiman using a temperature range of 19° – 30 °C. A number of neurons were held long enough to effect a temperature change of 5 – 10 °C. Again, in all neurons, the CF was lower at lower temperatures. The magnitude of the shift varied between fibers from an extreme high of 0.67 octave/6.5 °C for a fiber of CF 715 Hz to a low of 0.28 octave/6 °C for a fiber of CF 1225 Hz. Any change in Q value for the tuning curves was small, but the spontaneous activity was reduced at lower temperatures. Thresholds at CF were hardly affected. Their average shift of about 0.06 octave/°C agrees (Fig. 63) with the data of Eatock and Manley (1981). The data of Fengler et al. (1978) suggest that the shift is proportional to CF, of the order of 6% of CF per °C.

Finally, Hartline (1971b), recording evoked potentials from the midbrain of snakes, found a temperature dependence of the sensitivity and best frequency re-

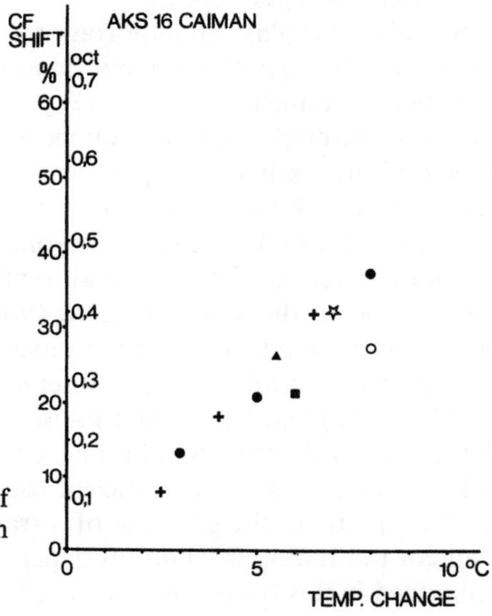

Fig. 63. Shift of CF of auditory nerve fibers of the caiman with temperature, in percent and in octaves. Each *symbol* represents one fiber. Data from one animal. From Klinke (1978)

sponse. At higher temperatures within the range used, the best auditory sensitivity was achieved at a higher frequency.

9 Discussion

Although a number of factors have been discussed in the above text, it is appropriate at this point to discuss in greater depth some selected topics which are important and of current interest.

9.1 Peripheral Filtering in the Reptilian Ear

Most peripheral filtering can be regarded as an irreversible loss of certain parts of the information present in the original signal. In the ear, a significant role in filtering seems to be played by the properties of those accessory structures which are interposed between the airborne sound and the receptor cells themselves. Certain features of these accessory structures are intrinsic and cannot be eliminated by adaptation, thus forming important restraints on the range and variety of signals which the organisms can use for communication. The evolution of the auditory periphery is to a large extent the history of changes and elaborations of these accessory structures. One significant trend is clear: peripheral filtering plays a highly important role in auditory analysis in amphibians and most reptiles, whereas in some reptiles and in birds and mammals the tendency has been to make the influence of such filtering as uniform as possible across all primary nerve fibers, such that these fibers have activity patterns which differ mainly in their CF. Further analysis is postponed and performed by correspondingly better-developed brain nuclei. This uniformity is not seen in the amphibia and most reptiles.

The middle ear plays an important role in limiting the overall range of frequency response, through the narrow band-pass characteristic of the eardrum-extracolumella-columella complex (Fig. 4) and probably a role in emphasizing laterality through the open connection between the middle ears. Thus, even at this level, not only are frequencies above 8 – 10 kHz filtered out, but the ability of a sound to initiate an inner-ear response probably depends strongly on where in space the sound originated. Beyond the middle ear, it is not clear whether the basilar membrane vibration in all reptiles depends as strongly on the middle ear response as in the alligator lizard (Weiss et al. 1978b) or whether the cochleas with a clear gradient of basilar membrane dimensions, as in geckos, have a traveling wave similar to that observed in the mammalian cochlea.

Klinke (1978) and Klinke and Pause (1979) argue that the fact that the response latencies of primary fibers in the caiman to clicks show a latency related to their CF, as in mammals, is insufficient foundation for the conclusion that the latency pattern is due to the presence of a traveling wave and express doubt that in the caiman the tuning mechanism depends on a traveling wave. In the gecko data described in this review the evidence indicates that the distribution of frequencies on the basilar papilla is probably exactly the reverse of what one would expect from simple observation of the physical tapering of the basilar papilla. In the alligator lizard, monitor lizard, and turtle a tonotopic organization of the papilla is present in the absence of clear morphological tapering of the papilla (Fig. 25) and in the alligator lizard with the basilar membrane motion more or less uniform over the entire membrane. In these cases a clear tonotopic organization is present without a traveling wave and must be based on other morphological gradients such as length of hair cell cilia and/or intrinsic tuning of the hair cells, as described in the turtle. It may at this time be relevant to generalize the comparative data and ask of the mammal ear to what extent the traveling wave and vibratory response of the basilar membrane are directly influenced by the filter properties of the complex hair cell system overlying it (see also, e.g., Kim et al. 1980). It is certainly plausible, indeed likely, that the total response of the system in all terrestrial vertebrates is a complex result of physical properties of the basilar membrane as such and of the properties of the hair cell-tectorial membrane complex overlying it, with interactive feedback depending on the physiological condition of the system. The fact that the vibration properties of the basilar membrane change within minutes after the death of the animal (squirrel monkey, Rhode 1973) supports the notion that the normal motion of the membrane depends on the physiological integrity of the organ of Corti.

Beyond the basilar membrane, the presence or absence and structure of the tectorial membrane can play a role in peripheral filtering. In the alligator lizard, Weiss et al. (1978b) feel that the striking difference in morphology of tectorial and ciliary structures between the apical and basal regions of the papilla form the basis of the dichotomous tonotopicity and that the gradation in length of the cilia in the basal region may form the basis of the graded tonotopic organization seen there. In the monitor lizard, both a dichotomous and a graded tonotopic organization of similar nature are seen, although here a thick tectorial membrane is present throughout. As it is not clear whether cilial length in the monitor papilla varies along its length, it is at the moment not possible to draw a complete parallel between these two species. Similarly, it is not clear if, and to what extent, the diversity of structures in the tectorial membrane of geckos plays a role in the variety of PSTH discharge patterns seen.

As is well known, most lizards are diurnal creatures which, in most cases, do not communicate vocally with each other. It is not clear to what extent the complexity of the peripheral filtering seen might have influenced the development of such vocal signals. It is, however, likely that the pronounced pulsed structure of most gecko calls, for example, have been optimized to produce a stimulus to which the ear responds extremely well. Thus the call structure in different species may be influenced by the individuality of the peripheral filtering arrangements, as long known in anuran amphibians (see, e.g., Caprancia 1976). It may be that the gecko-type PSTH pattern, with high synchrony of the first spike(s) in repeated response bursts may be partly due to the form of the tectorial structures. Similar PSTH were seen in the CN of the skink *Mabuya*, which also has such tectorial structures.

Holton and Weiss (1978) indicate that two-tone suppression is not observed in the higher CF group of nerve fibers from the alligator lizard papilla, whereas the apical fibers do exibit this phenomenon. This would support the notion that the tectorial membrane plays a key role in this phenomenon, as the tectorial membrane is absent from the basal region. Other mechanisms cannot, however, be excluded at the present time. The phenomenon may have a similar basis as the peripheral suppression seen in the ear of amphibians (Capranica and Moffat 1974). In gecko fibers and monitor lizard fibers (Manley, unpublished data) this phenomenon is observed, as it was in caiman fibers (Klinke 1978; Klinke and Pause 1979). In each case, tectorial material is present throughout.

The long list of possibilities for peripheral filtering of the auditory input ends with the hair cell properties, arrangement, and innervation patterns. The innervation patterns are virtually unknown and it is not clear which hair cells share a common innervation. Indirect evidence exists (Manley 1977) which indicates that nerve fibers in the monitor lizard innervate hair cells of opposing orientations and that this possibly influences the PSTH shape. What real significance this has, however, is unclear, as is also the reason for the consistent presence of a unidirectional area and why this area should be responsible for the low-frequency response.

9.2 Electrical Tuning of Hair Cells

Apart from the influence of the cilial length on the hair cells, as discussed above, at least for the case where the tectorial membrane is not present, other properties of hair cells are important in peripheral filtering, at least as far as frequency tuning is concerned. The most direct evidence for an influence of hair cell electrical properties on their frequency response comes from the previously described work of Fettiplace and Crawford (1978) and Crawford and Fettiplace (1978, 1979). Thus, turtle hair cells show a pronounced tendency under the influence of injected current pulses and loud low-frequency sound stimulation, to "ring" at frequencies centered around the CF (Fig. 15B). Fettiplace and Crawford state: "It is conceivable that an electrical tuning of the hair cells, similar to that proposed here for the terrapin cochlea may also operate in the mammal at least for frequencies less than about 1 kHz." Other effects discussed in this review, which may also be related to an intrinsic tuning mechanism are, e.g., on-off discharge patterns of primary fibers, preferred intervals in spontaneous discharge, and higher high-frequency slopes of the tuning curves below CF 700 Hz than for somewhat higher CFs. That these effects are not seen in mammals speaks, however, against their notion. That the Q values and especially the slope of the low-frequency side of the tuning curves of primary fibers are not smoothly monotonically related to the CF in the gecko (Figs. 29 and 30) and starling (Manley and Leppelsack 1977 and unpublished data) has already indicated that the tuning mechanisms of high CF and lower CF (below about 1 kHz) might be due to a different balance of factors. In addition, the preferred intervals in the spontaneous activity of low CF fibers can be explained by the presence of an intrinsic electrical filter in the hair cells (Manley 1979). Crawford and Fettiplace (1980) have observed spontaneous fluctuations in hair cell voltage whose dominant frequency component is that of the CF of the cell, again in the complete absence of sound stimuli.

It is therefore apparent that the frequency tuning properties of reptilian hair cells are due to the combination of a broad filtering from the middle ear and basilar membrane, refined by sharpening from mechanical influences from the cilia alone or cilia and tectorial membrane together and by the electrical properties of the hair cells themselves. Also, because the Q values of reptilian low-frequency tuning curves tend to be higher than those of mammals, it may be that the electrical influence at low frequencies is not seen in the mammalian cochlea, sharpening there relying on cilial-tectorial membrane effects and electrical-mechanical feedback influences, with the outer hair cells playing a key role in influencing the inner hair cells, with which most of the nerve fibers are connected (Manley 1978). Although a tectorial membrane is present through the whole mammalian cochlea, the inner and outer hair cells' cilia may not be equally firmly attached (Manley 1978). The cilia of both inner and outer hair cells increase in length toward the apex (Lim 1979), as also shown in *Caiman* and *Gerrhonotus*, and it is possible that cilial length plays an important role in frequency tuning in all vertebrate cochleae.

It is disturbing that the hair cell intracellular data from the alligator lizard (Mulroy et al. 1974) show significant differences from the turtle data. In particular, the magnitude of the oscillatory receptor potential [3 mV in the lizard compared to up to 40 mV in the turtle (cf. Fig. 15A)] is very different. In addition, Mulroy et al. report a significant DC component of the receptor potential, as seen at much higher frequencies in the guinea pig hair cell (Russell and Sellick 1977), whereas such a component was small except at high frequencies in the turtle hair cells (Crawford and Fettiplace 1980). Only further work will indicate to what extent these variations can be directly attributed to species' differences.

9.3 Temperature Effects on Tuning

The temperature effect data discussed above for both CM and auditory nerve units indicate that the frequency sensitivity of hair cells and their afferent fibers shifts to higher frequencies at higher temperatures, and vice versa. This, of course, implies a shift of the tonotopic organization of auditory centers throughout the brain. The data at this stage do not clearly indicate whether the shift in CF per °C is CF dependent or independent. The data of Smolders and Klinke (1977) for high CF cat auditory nerve fibers show only a small shift with temperature (less than 0.01 octave/°C), which would tend to suggest, if a common mechanism is involved, that a constant CF shift independent of CF is more likely to be the case. The failure to find a temperature effect on the absolute pitch determinations of human subjects at low frequencies (Emde and Klinke 1977) may be important in this regard. However, as the mechanisms of absolute pitch perception probably involve a series of neuronal interactions at stages higher than the auditory nerve, these data may not be conclusive.

It is of considerable interest that the tuning in the frequency domain of certain electroreceptor cells of weakly electric fish (Hopkins 1976) is also temperature sensitive, because these receptor cells are phylogenetically related to the hair cells of the acousticolateralis system. For electroreceptor afferent fibers with CFs near 130 Hz, a 2 °C increment of temperature reversibly raised a fiber's CF by 10 – 30 Hz. These electroreceptors show large membrane potential oscillations when the electrical loading of the cell is reduced (e.g., by removing the water over the skin). These oscillations in gymnotid fish occur at the frequency of the electric organ discharge (Bennet 1967), to which frequency the electroreceptors are most sensitive (Hopkins 1976). Hopkins discusses the possibility that electroreceptor tuning is dependent on the membrane filter properties responsible for these oscillations and that temperature might affect the frequency of these oscillations and hence tuning. Thus a strong parallel exists to the suggested electrical tuning of hair cells in reptiles as discussed above. If this is the basis of the temperature effects, and if, as Fettiplace and Crawford (1978) suggest, this electrical tuning is mainly limited to low frequencies, then the small effect in high-CF cat nerve fibers is not surprising.

In the anuran amphibians, temperature effects have been demonstrated in auditory nerve fibers from the *amphibian* papilla of the toad *Bufo americanus* (Moffat and Capranica 1976), and resemble the effects found in the gecko and caiman. In the toad's basilar papilla, however, temperature shifts up to 10°C have no effect on tuning. This profound difference may be related to the fact that the basilar papilla is a simply tuned, mechanical, resonant structure (Capranica and Moffat 1977). In a given individual animal, CFs from the basilar papilla are all virtually the same (Capranica, personal communication). If the temperature effect operates via an influence on the electrical tuning properties of the cell membrane, it may not be surprising that a receptor with predominantly mechanical tuning is insensitive to temperature. Similar considerations apply to temperature effect data in the mammalian cochlea.

9.4 On-Off PSTH Patterns

On-off responses have also been found in the primary afferents of the green frog (Sachs 1964), a bat (Suga et al. 1975), the pigeon (Gross and Anderson 1974), and the starling (Manley and Leppelsack, unpublished data). In the bat, the on-off units are only found in units with CFs near the constant frequency component of the orientation pulse. A mechanical resonance of this area of the basilar membrane, revealed in cochlear microphonic data, provides a simple explanation for these results (Suga et al. 1975). Such a mechanism is not very likely in the other species due to the lack of special morphological adaptations. Since these effects are more common at off-CF frequencies and at higher intensities, the "ringing" phenomenon reported in turtle hair cells by Fettiplace and Crawford (1978) could provide the basis for a possible explanation. If based on an electrical tuning of hair cells, it may be possible to explain why the effect has never been observed in fibers from normal mammalian cochleas. The whole phenomenon of on-off discharge in primary fibers needs further detailed study before any effective conclusions can be reached.

10 Conclusions with Regard to Further Work

While the study of auditory physiology in reptiles is clearly still rising from its infancy, a survey of the accumulated material clearly indicates its value in our general understanding of auditory systems. The unique features of the reptilian auditory system – including its systematic structural diversity – are at the same time a temptation and a challenge to the research worker. The material can certainly provide the basis for answering quite directed questions. The challenge lies in being able to deduce which features of the response can be truly generalized to other groups, such as the mammals. I believe this review indicates that we have learned a great deal in the past 10 years, but also shows the gaps which still exist in our knowledge. These gaps have been pointed out in some parts of the review as areas requiring more work. As always, the physiologist leans rather heavily on the interest and productivity of the anatomist. While great progress has been made in our understanding of the structure of the reptilian ear, we still know virtually nothing about important details concerning the innervation patterns. This lack is already a hindrance to progress in this field. Although it is true to say that the largest gaps in our knowledge concern the higher auditory centers, current interest and debate will continue to spur on the study of the peripheral auditory system. It is already clear that such comparative data will provide an essential contribution to the eventual understanding of the function of the mammalian ear.

Acknowledgments. I thank those colleagues who have allowed me to use joint unpublished data and to reproduce figures, and R. Klinke, H.J. Leppelsack, J. Manley, M.R. Miller, and E. Zwicker for valuable comments on an earlier version of the manuscript.

References

Adrian ED, Craik KJW, Sturdy RS (1938) The electrical response of the auditory mechanism in cold-blooded vertebrates. Proc R Soc Lond [Biol] 125:435 – 455

Angelborg C, Engström H (1973) The normal organ of Corti. In: Møller AR (ed) Basic mechanisms in hearing. Academic Press, New York London, pp 125 – 182

Ariens Kappers CV, Huber CG, Crosby EC (1960) The comparative anatomy of the nervous system of vertebrates, including man. Hafner Press, New York

Bagger-Sjöbäck D (1976) The cellular organization and neuronal supply of the basilar papilla in the lizard *Calotes versicolor*. Cell Tissue Res 165:141 – 156

Bagger-Sjöbäck D, Flock A (1977) Freeze-fracturing of the auditory basilar papilla in the lizard *Calotes versicolor*. Cell Tissue Res 177:431 – 443

Bagger-Sjöbäck D, Wersäll J (1973) The sensory hairs and tectorial membrane of the basilar papilla in the lizard *Calotes versicolor*. J Neurocytol 2:329 – 350

Baird IL (1960) A survey of the periotic labyrinth in some representative recent reptiles. Univ Kans Sci Bull 41:891 – 981

Baird IL (1970) The anatomy of the reptilian ear. In: Gans C, Parsons TS (eds) Biology of the reptilia, Vol 2. Academic Press, London New York, pp 193 – 275

Baird IL (1976) Anatomical features of the inner ear in submammalian vertebrates. In: Keidel WD, Neff WD (eds) Anatomy, physiology (ear). Springer, Berlin Heidelberg New York (Handbook of sensory physiology, vol V/1, pp 159 – 212)

Bennet M (1967) Mechanisms of electroreception. In: Cahn P (ed) *Lateral line detectors.* Univ. Indiana Press, Bloomington, pp 313 – 393

Berman DS, Regal PJ (1967) The loss of the ophidian middle ear. Evolution 21:641 – 643
Boord RL (1969) The anatomy of the avian auditory system. Ann N Y Acad Sci 167:186 – 198
Browner R, Caspary D (1976) The neurobiology of the acoustic tubercle in *Tupinambis nigropunctatus*. Society Neurosciences 6th Annal Meeting [Abstr] 2:(1)15
Campbell HW (1969) The effects of temperature on the auditory sensitivity of lizards. Physiol Zool 42:183 – 210
Campbell CBG, Boord RL (1974) Central auditory pathways of non-mammalian vertebrates. In: Keidel WD, Neff WD (eds) Anatomy, physiology (ear). Springer, Berlin Heidelberg New York (Handbook of sensory physiology vol V/1, pp 337 – 362)
Capranica RR (1976) Morphology and physiology of the auditory system. In: Llinas R, Precht W (eds) Frog neurobiology. Springer, Berlin Heidelberg New York
Capranica RR, Moffat AJ (1974) Evidence for mechanical origin of peripheral inhibition in the anuran inner ear. J Acoust Soc Am 55:S85
Carroll RL (1969) Problems of the origin of reptiles. Biol Rev 44:393 – 432
Coles RB, Lewis DB, Hill KG, Hutchins ME, Gower DM (1980) Directional hearing in the Japanese quail II Cochlear physiology. J Exp Biol 86:153 – 170
Crawford AC, Fettiplace R (1978) Ringing responses in cochlear hair cells of the turtle. J Physiol (Lond) 284:120 – 122
Crawford AC, Fettiplace R (1979) Reversal of hair cell responses by current. J Physiol (Lond) 295:66P
Crawford AC, Fettiplace R (1980) The frequency selectivity of auditory nerve fibers and hair cells in the cochlea of the turtle. J Physiol (Lond) 306:79 – 126
Davis H (1968) Mechanisms of the inner ear. Ann Otol Rhinol Laryngol 77:644 – 655
von Düring M, Karduck A, Richter H-G (1974) The fine structure of the inner ear in *Caiman crocodilus*. Z Anat Entwickl-Gesch 145:41 – 65
Eatock RA (1978) Auditory-nerve fibre activity in the Tokay gecko. M. Sc. Thesis, McGill University, Montreal
Eatock RA, Manley GA (1976) Temperature effects on single auditory-nerve fibre responses. J Acoust Soc Am 60:S80
Eatock RA, Manley GA (1981) Auditory nerve fibre activity in the Tokay Gecko: II, Temperature effect on tuning. J Comp Physiol [A] 142:219 – 226
Eatock RA, Manley GA, Pawson L (1981) Auditory nerve fibre activity in the Tokay Gecko: I, Implications for cochlear processing. J Comp Physiol [A] 142:203 – 218
Emde C, Klinke R (1977) Does absolute pitch depend on an internal clock? INSERM 68:145
Engström H (1967) The ultrastructure of the sensory cells of the cochlea. J Larygnol Otol 81:687 – 715
Fengler R, Klinke R, Pause M, Smolders H (1978) Reverse correlation in primary auditory nerve fibres of the caiman. Pfluegers Archiv [Suppl] 373:R85
Fettiplace R, Crawford AC (1978) The coding of sound pressure and frequency in cochlear hair cells of the terrapin. Proc R Soc Lond [Biol] 203:209 – 218
Flock A, Jørgensen M, Russel I (1973) The physiology of individual hair cells and their synapses. In: Møller AR (ed) Basic mechanisms in hearing. Academic Press, New York London
Foster RE, Hall WC (1978) The organization of central auditory pathways in a reptile, *Iguana iguana*. J Comp Neurol 178:783 – 832
Frankenberg E (1974) Vocalization of males of three geographical forms of *Ptyodactylus* from Israel. J Herpetol 8:59 – 70
Glatt AF (1975a) Vergleichend morphologische Untersuchungen am akustischen System einiger ausgewählter Reptilien. A *Caiman crocodilus*. Rev Suisse Zool 82:257 – 281
Glatt AF (1975b) Vergleichend morphologische Untersuchungen am akustischen System einiger ausgewählter Reptilien. B. Sauria, Testudines. Rev Suisse Zool 82:469 – 494

Göttl KH, Klinke R (1977) Differential susceptibility of positive- and negative- going portion of microphonics to anoxia. INSERM 68:103–104
Gross NB, Anderson DJ (1976) Single unit responses recorded from the first-order neuron of the pigeon auditory system. Brain Res 101:209–222
Guinan JJ, Guinan SI, Norris BM (1972) Single auditory units in the superior olivary complex; 1. Responses to sounds and classifications based on physiological properties. Int J Neurosci 4:101–120
Hartline P (1971a) Physiological basis for detection of sound and vibration in snakes. J Exp Biol 54:349–371
Hartline P (1971b) Mid-brain responses of the auditory and somatic vibration systems in snakes. J Exp Biol 54:373–390
Hartline P, Campbell HW (1969) Auditory and vibratory responses in the midbrains of snakes. Science 163:1221–1223
Hepp-Reymond M-L, Palin J (1968) Patterns in the cochlear potentials of the Tokay gecko (*Gekko Gecko*). Acta Otolarygnol 65:270–292
Hill KG, Boyan GS (1976) Directional hearing in crickets. Nature 262:390–391
Hill KG, Boyan GS (1977) Sensitivity to frequency and direction of sound in the auditory system of crickets. J Comp Physiol [A] 121:79–97
Hill KG, Lewis DB, Hitchings ME, Coles RB (1980) Directional hearing in the Japanese Quail, I.: Acoustic properties of the auditory systems. J Exp Biol 86: 135–151
Holton T, Weiss TF (1978) Two-tone rate suppression in lizard cochlear-nerve fibres: relation to receptor organ morphology. Brain Res 159:219–222
Hopkins CD (1976) Stimulus filtering and electroreception: tuberous electroreceptors in three species of gymnotid fish. J Comp Physiol [A] 111:171–207
Johnstone JR, Johnstone BM (1972) Unit responses from the lizard auditory nerve. Exptl Neurol 24:528–537
Johnstone BM, Sellick PM (1972) The peripheral auditory apparatus Q Rev Biophys 5:1–57
Johnstone BM, Taylor KJ (1971) Physiology of the middle-ear transmission system. J Otolaryngol Soc Australia 3:226–228
Johnstone CG, Schmidt RS, Johnstone BM (1963) Sodium and potassium in vertebrate cochlear endolymph as determined by flame microspectrophotometry. Comp Biochem Physiol 9:335–341
Kauffmann G (1974) Zur Abhängigkeit der Cochleapotentiale des Kaimans vom Stoffwechsel, von aktiven Transporten und von der Temperatur. J Comp Physiol [A] 90:245–273
Kauffmann G, Schwartzkopff J (1971) On the dependence on metabolism of the cochlear potentials in the caiman. Z Vergl Physiol 75:105–107
Kennedy MC (1974) Auditory multiple-unit activity in the midbrain of the Tokay gecko (*Gekko gecko* L.) Brain Behav Evol 10:257–264
Kennedy MC (1975) Vocalization elicited in a lizard by electrical stimulation of the midbrain. Brain Res 91:321–325
Kim DO, Molnar CE, Matthews JW (1980) Cochlear mechanics: Physiologically vulnerable nonlinear behaviour as reflected in ear-canal pressure and cochlear-nerve-fibre responses. J Acoust Soc Am 67:1704–1721
Klinke R (1978) Frequency analysis in the inner ear of mammals in comparison to other vertebrates. Verh Dtsch Zool Ges 1978:1–15
Klinke R, Pause M (1980) Discharge properties of primary auditory fibres in the caiman, *Caiman crocodilus:* comparisons and contrasts to the mammalian auditory nerve. Exp Brain Res 38, 137–150
Klinke R, Smolders J (1977) The performance of a primitive hearing organ of the cochlea type: primary fibre studies in the caiman. Addendum. Effect of temperature shift on tuning properties. In: Evans EF, Wilson JP (eds) Psychophysics and physiology of hearing. Academic Press, London

Konishi M (1970) Comparative neurophysiological studies of hearing and vocalizations in songbirds. Z Vergl Physiol 66:257 – 272
Knudsen EI, Konishi M (1978) Space and frequency are represented separately in auditory midbrain of the owl. J Neurophysiol 41:870 – 884
Leake PA (1974) Central projections of the statoacoustic nerve in *Caiman crocodilus.* Brain Behav Evol 10:170 – 196
Leake PA (1976) Scanning electron microscopy of labyrinthine sensory organs in *Caiman crocodilus.* Scan Electron Microsc II:277 – 284
Leake PA (1977) SEM observations of the cochlear duct in *Caiman crocodilus.* Scan Electron Microsc II:437 – 444
Leppelsack H-J (1974) Funktionelle Eigenschaften der Hörbahn im Feld L des Neostriatum caudale des Staren. J Comp Physiol [A] 88:271 – 320
Lim DJ (1979) Cochlear anatomy related to cochlear mechanics. J Acoust Soc Am 65:S27
Manley GA (1970a) Frequency sensitivity of auditory neurons in the caiman cochlear nucleus. Z Vergl Physiol 66:251 – 256
Manley GA (1970b) Comparative studies of auditory physiology in reptiles. Z Vergl Physiol 67:363 – 382
Manley GA (1971) Some aspects of the evolution of hearing in vertebrates. Nature (Lond) 230:506 – 509
Manley GA (1972a) The middle ear of the Tokay gecko. J Comp Physiol [A] 81:239 – 250
Manley GA (1972b) Frequency response of the middle ear of geckos. J Comp Physiol [A] 81:251 – 258
Manley GA (1972c) Frequency response of the ear of the Tokay gecko. J Exp Zool 181:159 – 168
Manley GA (1973) A review of some current concepts of the functional evolution of the ear in terrestrial vertebrates. Evolution 26:608 – 621
Manley GA (1974) Activity patterns in the peripheral auditory system of some reptiles. Brain Behav Evolution 10:244 – 256
Manley GA (1976) Auditory responses from the medulla of the monitor lizard *Varanus bengalensis.* Brain Res 102:329 – 334
Manley GA (1977) Response patterns and peripheral origin of auditory nerve fibers in the monitor lizard, *Varanus bengalensis.* J Comp Physiol (A) 118:249 – 260
Manley GA (1978) Cochlear frequency sharpening – a new synthesis. Acta Otolaryngol 85:167 – 176
Manley GA (1979) Preferred intervals in the spontaneous activity of primary auditory neurons. Naturwissenschaften 66:582 – 583
Manley GA, Johnstone BM (1974) Middle ear function in the Guinea Pig. J Acoust Soc Amer 56:571 – 576
Manley GA, Leppelsack H-J (1977) Preliminary data on activity patterns of cochlear ganglion cells in the starling, *Sturnus vulgaris.* INSERM 68:127 – 136
Manley GA, Robertson D (1976) Analysis of spontaneous activity of auditory neurones in the spiral ganglion of the guinea pig cochlea. J Physiol (Lond) 258:323 – 336
Manley GA, Irvine D, Johnstone BM (1972) Frequency response of the bat tympanic membrane. Nature (Lond) 237:112 – 113
Manley J (1971) Single unit studies in the midbrain auditory area of *Caiman.* Z Vergl Physiol 71:255 – 261
Marcellini D (1977) Acoustic and visual display behaviour of gekkonid lizards. Am Zool 17:251 – 260
McGill TE (1960) A review of hearing in amphibians and reptiles. Psychol Bull 57:165 – 168
Miller MR (1966a) The cochlear duct of lizards. Proc Calif Acad Sci 33:255 – 359
Miller MR (1966b) The cochlear ducts of lizards and snakes. Am Zool 6:421 – 429
Miller MR (1968) The cochlear duct of snakes. Proc Calif Acad Sci 35:425 – 576

Miller MR (1973a) A scanning electron microscope study of the papilla basilaris of *Gekko gecko*. Z Zellforsch 136:307–328

Miller MR (1973b) Scanning electron microscope studies of some lizard basilar papillae. Am J Anat 138:301–330

Miller MR (1974) Scanning electron microscope studies of some skink Papillae basilares. Cell Tiss Res 150:125–141

Miller MR (1975) The cochlear nuclei of lizards. J Comp Neurol 159:375–406

Miller MR (1978a) Further scanning electron microscope studies of lizard auditory papillae. J Morphol 156:381–418

Miller MR (1978b) Scanning electron microscope studies of the papilla basilaris of some turtles and snakes. Am J Anat 151:409–436

Miller MR (1980) The Reptilian cochlear duct. In: Popper AN, Fay RR (eds) Comparative studies of hearing in vertebrates. Springer, Berlin Heidelberg New York

Miller MR, Kasahara M (1979) The cochlear nuclei of some turtles. J Comp Neurol 185:221–236

Moffat AJM, Capranica RR (1976) Effects of temperature on the response properties of auditory nerve fibres in the American Toad (*Bufo bufo*). J Acoust Soc Am 60:S80

Moffat AJM, Capranica RR (1978) Middle-ear sensitivity in anurans and reptiles measured by light-scattering spectroscopy. J Comp Physiol [A] 127:97–107

Mulroy MJ, Altmann DW, Weiss TF, Peake WT (1974) Intracellular electric responses to sound in a vertebrate cochlea. Nature 249:482–485

Nadol JB, Mulroy JJ, Goodenough DA, Weiss TF (1976) Tight and gap junctions in a vertebrate inner ear. Am J Anat 147:281–302

Necker R (1970) Zur Entstehung der Cochleapotentiale von Vögeln: Verhalten bei O_2-Mangel, Cyanidvergiftung und Unterkühlung sowie Beobachtungen über die räumliche Verteilung. Z Vergl Physiol 69:367–425

Olson EC (1966) The middle ear – morphological types in amphibians and reptiles. Am Zool 6:399–419

Page CH (1970) Electrophysiological study of auditory responses in the goldfish brain. J Neurophysiol 33:116–128

Paton JA, Moffat AJM, Capranica RR (1976) Electrophysiological correlates of basilar membrane motion in the turtle. J Acoust Soc Am 59:S46

Patterson WC (1966) Hearing in the turtle. J Aud Res 6:453–464

Peterson SK, Frischkopff LS, Lechène C, Oman CM, Weiss TF (1978) Element composition of inner ear lymphs in cats, lizards and skates determined by electron probe microanalysis of liquid samples. J Comp Physiol [A] 126:1–14

Pfeiffer RR (1966) Classification of response patterns of spike discharges for units in the cochlear nucleus; toneburst stimulation. Exp Brain Res 1:220–235

Potash LM (1970) Neuroanatomical regions relevant to production and analysis of vocalization within the avian torus semicircularis. Experientia 26:1104–1105

Potter HD (1965) Mesencephalic auditory region of the bullfrog. J Neurophysiol 28:1132–1154

Pritz MB (1974a) Ascending connections of a midbrain auditory area in a crocodile, *Caiman crocodilus*. J Comp Neurol 153:179–198

Pritz MB (1974b) Ascending connections of a thalamic auditory area in a crocodile, *Caiman crocodilus*. J Comp Neurol 153:199–214

Rhode WS (1973) An investigation of post-mortem cochlear mechanics using the Mössbauer effect. In: Møller AR, Boston P (eds) Basic mechanisms in hearing. Academic Press, New York London, pp 49–63

Robertson D (1976) Correspondence between sharp tuning and two-tone inhibition in primary auditory neurones. Nature 259:477–478

Robertson D, Manley GA (1974) Manipulation of frequency analysis in the cochlear ganglion of the guinea pig. J Comp Physiol [A] 91:363–375

Ridgeway SH, Wever EG, McCormick JG, Palin J, Anderson JH (1969) Hearing in the giant sea turtle, *Chelonia mydas.* Proc Natl Acad Sci 64:884 – 890
Russell IJ, Sellick PM (1977) Tuning properties of cochlear hair cells. Nature 267:858 – 860
Sachs MB, Kiang NY-S (1968) Two-tone inhibition in auditory-nerve fibers. J Acoust Soc Am 43:1120 – 1128
Sachs MB, Sinnott JM (1978) Responses to tones of single cells in Nucleus Magnocellularis and Nucleus Angularis of the Redwing Blackbird (Agelaius phoenicus). J Comp Physiol [A] 126:347 – 361
Sachs MB, Young ED, Lewis RH (1974) Discharge patterns of single fibers in the pigeon auditory nerve. Brain Res 70:431 – 447
Sammaritano-Klein MR (1976) Single-unit responses in the midbrain auditory nucleus of the lizard *Gekko gecko.* Master's thesis, McGill University, Montreal
Sammaritano-Klein MR, Manley GA (1976) Auditory responses of single neurons in the midbrain of the Tokay gecko. J Acoust Soc Am 59:S46
Smith CA (1968) Ultrastructure of the organ of Corti. Advancement of Science June 1968:419 – 433
Smolders J, Klinke R (1977) Effect of temperature changes on tuning properties of primary auditory fibres in caiman and cat. INSERM 68:125 – 126
Suga N, Campbell HW (1967) Frequency sensitivity of single auditory neurons in the gecko *Coleonyx variegatus.* Science 157:88 – 90
Suga N, Simmons JA, Jen PH-S (1975) Peripheral specialization for fine analysis of doppler-shifted echos in the auditory system of the 'CF-FM' bat, *Pteronotus parnellii.* J Exp Biol 63:161 – 192
Takasaka T, Smith CA (1971) The structure and innervation of the pigeon's basilar papilla. J Ultrastruct Res 35:20 – 65
Walsh BT, Miller JB, Gacek RR, Kiang NY-S (1972) Spontaneous activity in the eighth cranial nerve of the cat. Int J Neurosci 3:221 – 236
Weisbach W, Schwartzkopff J (1967) Nervöse Antworten auf Schallreiz im Gehirn von Krokodilen. Naturwissenschaften 24:650
Weiss TF, Mulroy MJ, Altmann DW (1974) Intracellular responses to acoustic clicks in the inner ear of the alligator lizard. J Acoust Soc Am 55:606 – 619
Weiss TF, Mulroy MJ, Turner RG, Pike LL (1976) Tuning of single fibres in the cochlear nerve of the alligator lizard: relation to receptor morphology. Brain Res 115:71 – 90
Weiss TF, Altmann DW, Mulroy MJ (1978a) Endolymphatic and Intracellular resting potential in the alligator lizard cochlea. Pflueger Arch 373:77 – 84
Weiss TF, Peake WT, Ling A, Holton T (1978b) Which structures determine frequency selectivity and tonotopic organization of vertebrate cochlear nerve fibers? Evidence from the alligator lizard. In: Naunton RF (ed) Evoked electrical activity in the auditory nervous system. Academic Press, London New York, pp 91 – 112
Werner YL (1972) Temperature effects on inner-ear-sensitivity in six species of iguanid lizards. J Herpetol 6:147 – 177
Werner YL (1976) Optimal temperatures for inner-ear performance in gekkonid lizards. J Exp Zool 195:319 – 352
Werner YL, Wever EG (1972) The function of the middle ear in lizards: *Gekko gecko* and *Eublepharis macularius (Gekkonoidea).* J Exp Zool 179:1 – 16
Wersäll J, Flock A (1967) Morphological aspects of cochlear hair cell physiology. In: Graham AB (ed) Sensorineural hearing processes and disorders. Little, Brown, Boston, pp 3 – 19
Wever EG (1978) The reptile ear. Princeton Univ. Press, Princeton, N. J.
Wever EG, Werner YL (1970) The function of the middle ear in lizards: *Crotaphytus collaris.* J Exp Zool 175:327 – 342

Recent Advances in Structural Correlates of Auditory Receptors

C. A. Smith *

University of Oregon Medical School, 3515 S.W. Veterans Hospital Road, Portland, OR 97201 (USA)

---

* The author is very grateful to Mrs. Nancy Schuff, who assisted with the bibliography and the illustrations, to Dr. Robert Brummett and Dr. Mary Meikle, who critically reviewed the manuscript, to Mrs. Suzanne Moody for the expert artwork, and to Miss Dorrine Conrad for repeated typings of the manuscript. The courtesy of Dr. E. Wever, Dr. Malcolm Miller, and Dr. Thomas Weiss in permitting reproduction of figures from their papers is gratefully acknowledged. The preparation of the manuscript as well as much of the personal research involved was supported by NIH research grant # NS 08813

1 Introduction

The first comprehensive studies on the comparative structure of the vertebrate ear were made by Gustav Retzius in 1884. It was a subject which apparently held great fascination for him, and his two beautifully illustrated volumes give evidence of his scholarly application to detail. His contemporaries and those who followed were more interested in the inner ear of man than that of lower animals and there was certainly, and still is, a great deal to be learned about the mammalian cochlea.
Within the past 2 decades there has been a renewed interest in the inner ears of lower vertebrates. Wever's physiological studies in the early 1950s on reptilian ears provided an important stimulus to new morphological investigations on submammalian vertebrates. Furthermore, at about that time, research on mammals seemed to have reached a plateau. Investigators hoped that detailed observations on lower animals whose ears served a similar function but had a different histological organization might yield some clues about basic mechanisms common to all auditory receptors. The answers to many questions are still obscure but much new data about structural components have been produced. Only the highlights of the last decade will be covered in this review.

2 Insects

Sensory receptors in insects have received a fair amount of attention over the years from invertebrate physiologists, but very little from investigators primarily interested in the auditory system. The insect auditory receptors (chordotonal organs) described to date are rather simple in structure in that they are composed only of neurons and supporting cells. But it seems clear that they are mechanoreceptors, sensitive to deformation. Although not all mechanoreceptors are similar in structure (the Pacinian corpuscles are remarkably different from the organ of Corti), it seems possible that those serving a common sensory modality could have some common structural features. And indeed they do.
One of the first studies on the ultrastructure of an insect vibration receptor was made by Gray in 1960. He studied the locust and found that its tympanic organ was composed of an aggregate of bipolar neurons which were attached peripherally to a tympanic membrane. The membrane was a sheet of the exoskeleton whose inner surface was in contact with a layer of attachment cells (Fig. 1). The latter covered the "scolopale" cap and upper part of the "scolopale" cell, which in turn enclosed the upper part of the neuronal dendrite. The remainder of the soma and the axon process was covered by other supporting cells. One of Gray's interesting findings was that the tip of the dendrite had become specialized and had the structure of a cilium. In general, a motile cilium is composed of a basal body located at the base of the cilium plus an axoneme containing microtubules arranged in nine sets of peripheral doublets and two central singlets. The cilium of the locust receptor contained nine peripheral doublets, but only one tubule of

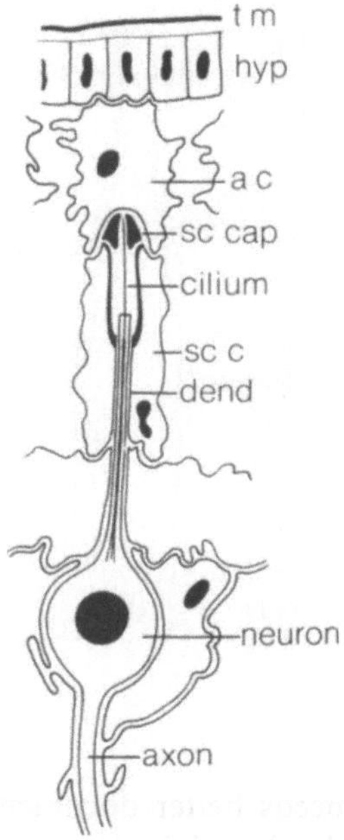

Fig. 1. Section through the auditory ganglion of the locust ear showing one sensory unit. *Ac*, attachment cell; *dend*, dendritic process; *hyp*, hypodermis; *sc c*, scolopale cell; *sc cap*, scolopale cap; *t m*, tympanic membrane. Redrawn from Gray (1960)

the doublet was hollow; the other appeared to be a solid rod. There were no central singlets. Since that time, cilia with basal bodies have been found in many sound receptors and other sensory receptors, as well.

Ghiradella (1971) identified cilia in the dendritic tips of the tympanic organ of the noctuid moth. She has given the internal structure as being 9 + 2 without further description. Young (1973) chose the chordotonal organ of the Australian cicada *(Cyclochila australasia)* as his subject. There were nine peripheral doublets which generally took the form of one rod and one tubule although on occasion both were hollow tubules. He found that dynein arms were present in the proximal ciliary shaft region only (Fig. 2). Two basal bodies were present, one at the base of the ciliary shaft and a second in the root process. The campaniform sensillum of the cockroach *(Blaberus discoidalis)* is another insect mechanoreceptor which includes a bipolar ganglion cell with a cilium (Moran et al. 1971). The cilium has a basal body from which the peripheral doublets (9 + 0) arise. No dynien arms have been described or illustrated. The internal character of the cilium changes quickly from basal body to peripheral tip. It becomes a thickened process, filled with several hundred microtubules and enclosed by a dense, extracellular material. The distal tip, which contains a dense, amorphous material surrounding the microtubules, inserts directly into the cuticular cap of the sensillum.

These mechanoreceptors have some morphological features in common. Firstly, a long cilium seems to be the actual receptor element but, in all cases where it has been clearly described and illustrated, the internal structure is somewhat different from that of motile cilia found in the respiratory tract. Both of the doublets may be tubules or one may be a solid rodlike structure. There are no central tubules either in the locust (Gray 1960), the cicada receptors (Young 1973), or in the campaniform sensilla of the cockroach (Moran et al. 1971). Ghiradella (1971) has stated that the cilium has a 9+2 pattern in the noctuid moth cells but this

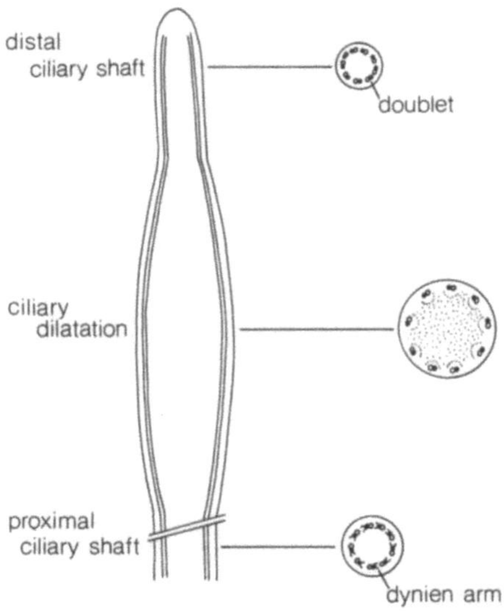

Fig. 2. The structure of the cilium on the auditory receptor cell of the cicada. Redrawn from Young (1973)

needs better documentation. The doublets of the locust cilia have some well-developed dynien arms but only in one part of the shaft (Young 1973). In short, the auditory cilia in the insect organs where they are well described and illustrated are similar to cilia known to be motile elsewhere, but they all show some differences in internal structure. This is not surprising as olfactory (Reese 1965) and retinal cilia likewise have structural adaptations (Sjöstrand 1956).

Secondly, the cap which covers the cilium and the parts of the supporting structure immediately surrounding it are composed of dense material which would lend a stiffness to the slender structure. This is common to scolopophorous cells. Another element lending stiffness is the multiplicity of microtubules found in the tip of the campaniform sensilla (Moran et al. 1971).

These two features, ciliary modification and apical stiffness, are prominent characteristics in the vertebrate ear also. Insect sensory cells differ from those of the acousticolateral line system of vertebrates in that the cilium is part of the *neuron* which is also the receptor cell. In that regard they are more like the receptors of the vertebrate olfactory system (Reese 1965).

The evidence from many investigators (summarized in Michelsen 1974) indicates that these insect sensory organs are sound receptors but that there is little or no frequency discrimination in a single organ. Katsuki and Suga (1960) studied the tympanal organs of several Orthoptera (Acridiudae, Tettigoniidae, and Gryllidae) and one Homoptera (Cicadidae) and found that the dominant frequency involved in stridulation produced by a particular insect coincided with the most sensitive part of its tympanal organ's frequency range. In the noctuid moth (Roeder and Treat 1961) sensitivity is best over a range that corresponds to the echo-locating chirps of their predator bats. Katsuki and Suga (1960) have also pointed out that there are vibration receptors other than the tympanal organs on the insect body and that it may be the neural patterns sent in to the central nervous system by some or several combinations of these that are most important to insects.

3 Reptiles

The vertebrate ear has some distinctive features which are not present in invertebrates. One outstanding characteristic is that the sensory receptors are specialized *epithelial cells* and that they have a similar cytological organization in all vertebrates. Each sensory cell is specialized by virtue of a cuticular plate on its apical end from which numerous hairs of variable, stepwise lengths protrude (Fig. 3). Most of the hairs (called stereocilia) are packed with microfilaments, not tubules. Only one cilium containing microtubules (the kinocilium) is present and in the mammalian cochleae there are no kinocilia at all. Usually the cilia are inserted into a gelatinous membrane. Nerve endings are present on the basal ends or on the sides of the cells.

These features are found in the lateral line organs, the vestibular labyrinthine receptor organs, the auditory papillae, and the cochleae. Receptor organs in the lateral line and vestibule are situated on firm connective tissue bases through which the nerve fibers and blood vessels pass. All vertebrate *vestibular* receptors, from fish to man, have this type of histological construction. On the other hand, the auditory receptor structures of reptiles, birds, and mammals are located on thin, flexible membranes. This membrane, called the basilar membrane, along with either papilla or the organ of Corti, separates the fluid of scala tympani from the special fluid which fills the cochlear duct.

The lateral line organs are exposed to salt or fresh water whereas the inner ear has its own special fluid environment. The closed epithelial tube which forms the membranous labyrinth is filled with one fluid named endolymph, which has a high potassium concentration. The outside of the tube is surrounded by a second fluid, called perilymph, which is much like extracellular tissue fluid elsewhere in that it has a low potassium and high sodium content.

The hair bundles on the sensory cells are composed of many stereocilia and one kinocilium (Fig. 3). The stereocilia which are filled with microfilaments are arranged in several rows in a graded or steplike pattern with the longest cilia on one side. The single kinocilium which contains microtubules is located in the center of and just outside the tallest row of stereocilia. As the kinocilium is always on one side of the bundle (Fig. 4), it is said that it "polarizes" the cell in a specific direction. Present evidence indicates that the polarization direction is related to activation of the sensory cell. Lowenstein and Wersäll (1959) and Flock (1965, 1971) showed that when the hair bundle was moved in a direction toward the kinocilium, excitation occurred; when it was moved in the opposite direction, inhibition took place. Groups of hair cells generally have a specific polarization direction in the reptilian ear. In other papillae, such as in the pigeon (Takasaka and Smith 1971) and chicken (Tanaka and Smith 1978), all sensory cells are polarized in the same general direction. The fully differentiated cochlear hair cells in mammals have only stereocilia and no kinocilia.

The membranous labyrinth of reptiles contains two sensory papillae anterior to the vestibule. These are the basilar papilla and the macula of the lagena. The lagenar macula is an otolith organ and persists in birds, but in mammals is found only in the monotremes (Smith and Takasaka 1971). Other living mammals do not have maculae in their cochlear ducts. The basilar papillae (the auditory recep-

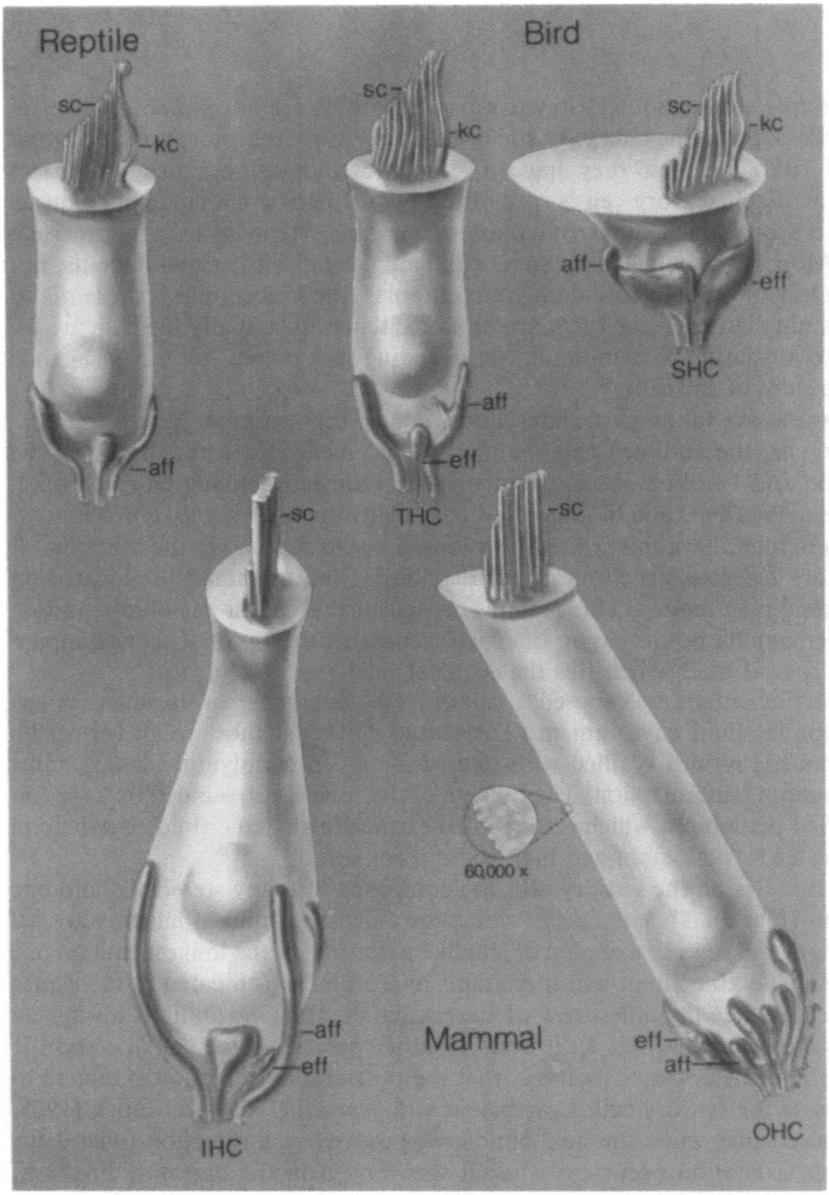

Fig. 3. Typical auditory hair cells from reptiles, birds and mammals. All are drawn to the same scale. *Aff*, afferent nerve ending; *eff*, efferent nerve ending; *IHC*, inner hair cell; *kc*, kinocilium; *OHC*, outer hair cell; *sc*, stereocilia; *SHC*, short hair cell; *THC*, tall hair cell. Drawn by Suzanne Moody

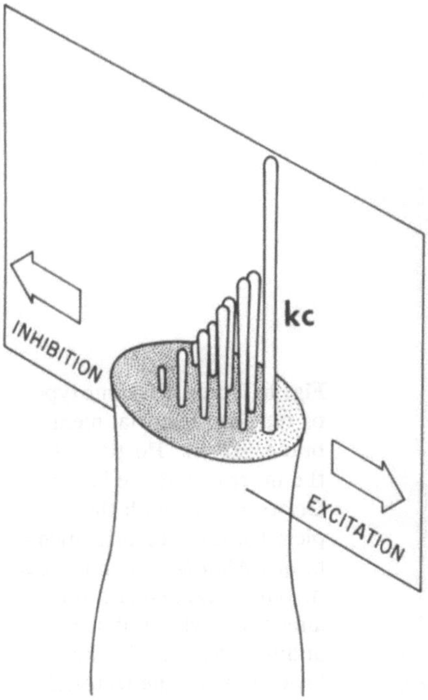

Fig. 4. Directional relationship between deflection of the hair bundle and activation of the hair cell. Bending toward the kinocilium (*kc*) results in excitation; bending away from the kinocilium results in inhibition. Redrawn from Flock. Reprinted from Smith (1975b)

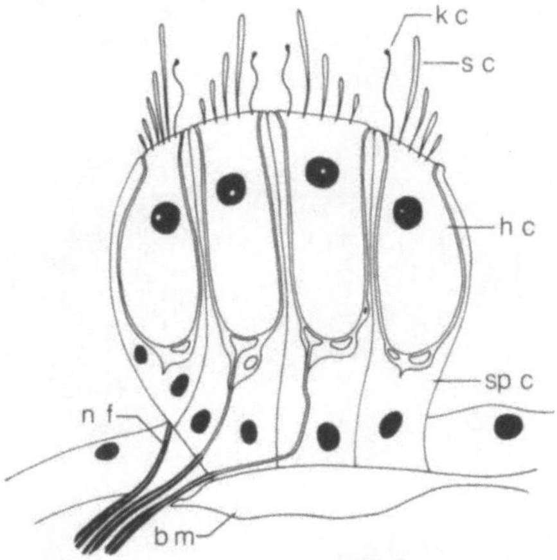

Fig. 5. Reptilian (alligator lizard) basilar papilla. The nerve fibers (*nf*) at lower left pierce the habenula perforata of the basilar membrane (*bm*) and terminate on the hair cells (*hc*). *Kc*, kinocilium; *sc*, stereocilium; *spc*, supporting cell. Redrawn from Mulroy (1974b)

tors) of all reptiles are quite small (Fig. 5). The lizard papillae, studied by Miller (1966), varied in length from 0.15 mm to 2.4 mm and in area from 0.013 mm^2 to 0.19 mm^2. In living reptiles, only the Crocodiliae have longer papillae: almost 4 mm in *Caiman crocodilus* and just over 5 mm in *Alligator mississippiensis*. They also have the largest numbers of sensory cells. *A. mississippiensis* has 10 800 hair cells (Wever 1978). Hair cell numbers in the lizard papillae vary from 50 (chameleons) to 1600 (*Gekko gekko*).

Miller pointed out that gross structure was fairly stable within any one family but microscopic examination has revealed many variations among species. There are differences in sensory cell number, kinociliar polarization, presence of "efferent" nerve fibers, and type and extent of the tectorial membrane. Some of these features were originally revealed by Wever, who has studied the histological arrangement and physiological properties of the ears of reptiles over the last 20 years and recently compiled these findings in a book (Wever 1978).

Wever's studies of the tectorial membranes of lizards (1967a, b) were made on sectioned material, and he demonstrated that the construction of the tectorial membrane varied from a fairly thick fibrous plate to a few dispersed filaments (Fig. 6). Remarkable variations are present even within a single individual.

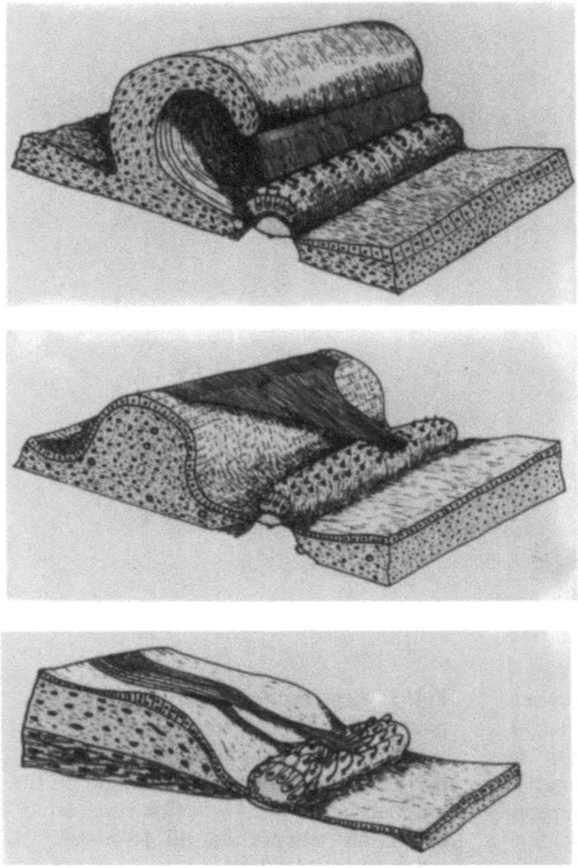

Fig. 6. Three different types of reptilian tectorial membrane. *Above:* Portion of the inner ear of the lizard *Gekko gecko*, with the complete form of tectorial membrane. *Middle:* The inner ear of *Iguana iguana* showing almost the whole of the auditory papilla. At the lower edge of the tectorial membrane is the short tectorial plate. *Below:* The inner ear of a chameleon, *Chamaeleo c. calcarifer*, showing the dendritic form of tectorial membrane. From Wever (1967b)

Miller's more recent studies, clarified by the three-dimensional view of scanning electron microscopy (1973a, b), have yielded interesting information. The tectorial membrane of *Eublepharis macularius*, for example, has regional variations. Miller's material clearly revealed the thickened masses of tectorial material (called "sallets" by Wever) which showed no connection to the limbus but which covered groups of hair cells and were attached to the tallest cilia. Other parts of the papilla had a very delicate type of membrane which was *attached* to the limbus. Wever (1967a) found the sallets in all the gekkonids. Another gekkonid, *Coleonyx variegatus*, has sallets in what Miller (1973a) calls the postaxial region (Fig. 7) whereas the neighboring sensory cells (preaxial) are covered by the more delicate filamentous, attached type of tectorial membrane. The basiliscines (Iguanidae) have a heavy tectorial membrane but this covers only a small number of the hair cells. Many other hair cells in this lizard's papilla seem to be covered by no tectorium at all (Wever 1967a; Miller 1973a). The tectorial membrane of *Caiman crocodilus* is thick and covers all the hair cells (Baird 1974). These brief descriptions illustrate only a few of the variations found in the relationship between the tectorial membrane and hair cells in reptiles. A total of six definite types of tectorial membrane attachments in lizards were described by Wever (1967a). He then added one more which did not seem to fit into such a classifica-

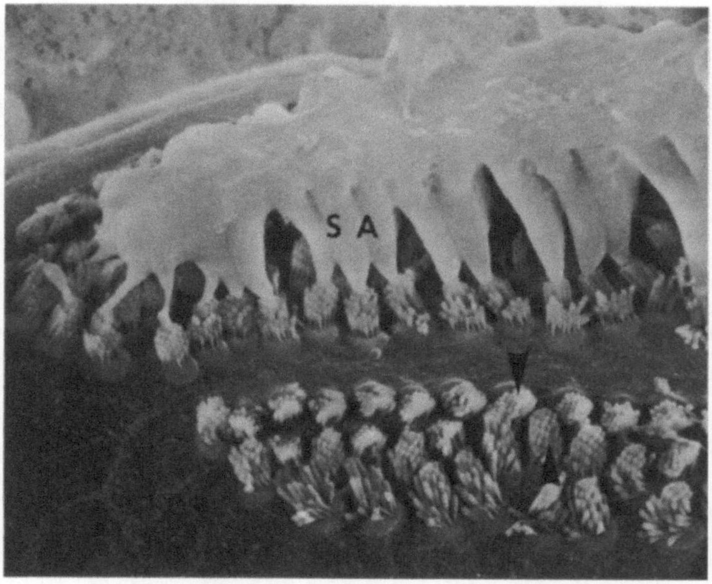

Fig. 7. Scanning electron micrograph (*SEM*) of the tip of the basilar papilla of *Coleonyx variegatus* illustrating the thick tectorial masses called sallets (*SA*). Note that the hair bundles at lower right have opposite polarizations (*arrow-heads*). ×1500. From Miller (1973a)

Fig. 8. SEM of hair bundles from *Coleonyx variegatus* showing the enlarged tip of each kinocilium (*KC*) and the stepwise lengths of the stereocilia (*SC*). The five hair cells are polarized in the same direction. *CP*, cuticular plate. ×4300. From Miller (1973a)

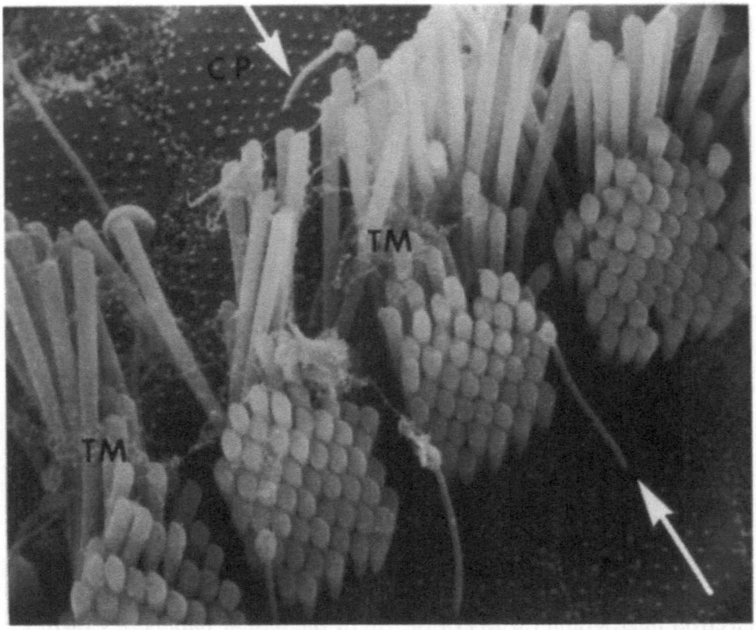

Fig. 9. SEM of four hair bundles from *Gerrhonotus multicarinatus*. Some filaments of tectorial membrane (*TM*) are visible around the hair cells. The deciliated cuticular plates (*CP*) of other hair cells are seen at *lower right* and *upper left*. Kinocilia with enlarged bulbs are still present on the deciliated hair cells and show opposing polarizations (*arrows*). ×4000. From Miller (1973a)

tion: a thin, loose network of filaments covering the hair cells, but with no other attachments.

Baird (1970), Miller (1973a, b), and Mulroy (1974b) found that each hair bundle contained a kinocilium but that the sensory cell polarization as indicated by the position of the kinocilium was not equivalent over the entire papilla (Figs. 7–9). The polarization direction often changed from proximal to distal end but the patterns thus produced were consistent within a species. There were also variations in the height of the ciliary bundle. In *Iguana iguana*, for example, hair cells in the central zone (Baird 1974) had short cilia whereas the cilia on hair cells on either end were twice as long.

Innervation patterns have not been studied in detail, but Baird (1974) found some groups of cells which had nerve endings filled with small vesicles and proposed that they were probably axon endings from cells other than bipolar cochlear neurons. Other cell groups had no nerve endings that could be identified as "efferent". Little is known about *afferent* nerve patterns in reptiles so that it is not clear what variations, if any, are present in sensory innervation.

Lizards show the greatest structural variability. Turtles, snakes, and amphisbaenians have less complex papillae and their ears seem to be more primitive from an evolutionary viewpoint (Baird 1974).

Even though the basilar membrane is freely suspended between endolymph and perilymph in the reptilian ear, the question arises as to whether a traveling waveform could be active on such a small papilla. Based on electrophysiological evidence, Manley (1970) inferred that some frequency analysis is present on the longer crocodilian papillae but made no speculations as to how this might be accomplished. Suga and Campbell (1967) measured neural responses to sound by use of micropipettes inserted into the medulla oblongata at regions they believed to be the nucleus magnocellularis dorsalis of the lizard, *Coleonyx variegatus*. The response curves obtained were not as sharply tuned as those in mammals and they concluded this might be due either to a lack of traveling wave, to specific innervation patterns, or to both.

Weiss, Mulroy, and their associates have attempted to clarify these phenomena and chose the ear of the alligator lizard (*Gerrhonotus multicarinatus*) for study because it has some unique structural features. The basilar papilla of this lizard is quite short (0.4 mm in length) and narrow. It has about 160 hair cells arranged in irregular transverse rows which contain four to six cells at the proximal end but only three cells at the distal tip. The entire papilla is remarkably convex so that the outermost cells on both sides are lower than the more central cells (Miller 1973a). It has several regional peculiarities which have been exploited by Weiss et al. (1978). The greatest structural variations occur between the basal (proximal) and the apical (distal) portions (Figs. 10, 11).

The *apical end* has a thick tectorial cap (Wever 1967a; Miller 1973a) which covers approximately 37 hair cells, about one-fourth of the total number (Miller 1973a). The cilia on these cells are all short, less than 10 µm long (Mulroy 1974a, b). There is one kinocilium in each hair bundle and the kinociliar orientation is unidirectional in all the cells covered by a tectorial membrane. Vesiculated nerve endings (probably efferent) are present on cells at the apical end.

The *basal* (*proximal*) three-fourths of the papilla has no organized tectorial membrane although some fibrous material covers the hair bundles (Wever 1967a; Miller 1973a; Mulroy 1974a, b). No efferent nerve endings have been identified. The cilia are long, increasing from 10 µm to almost 30 µm toward the basal end

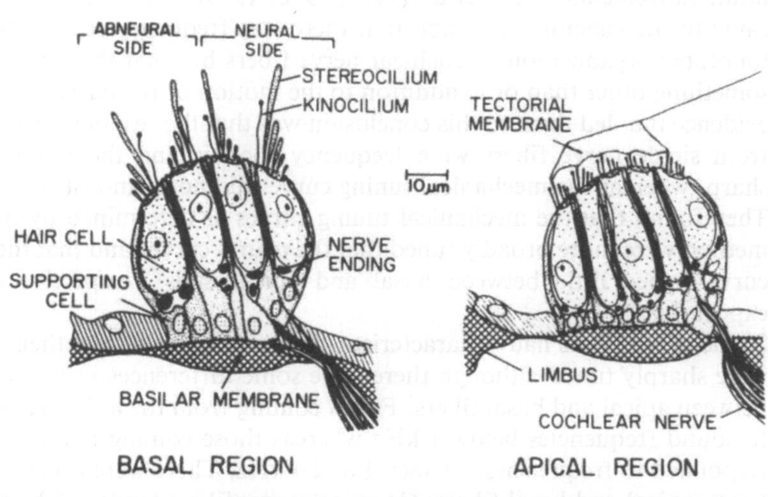

Fig. 10. Sections through the basal (proximal) and apical (distal) ends of the alligator lizard showing differences in tectorial membrane, in length of cilia, and in polarization of the kinocilia. From Weiss et al. (1978)

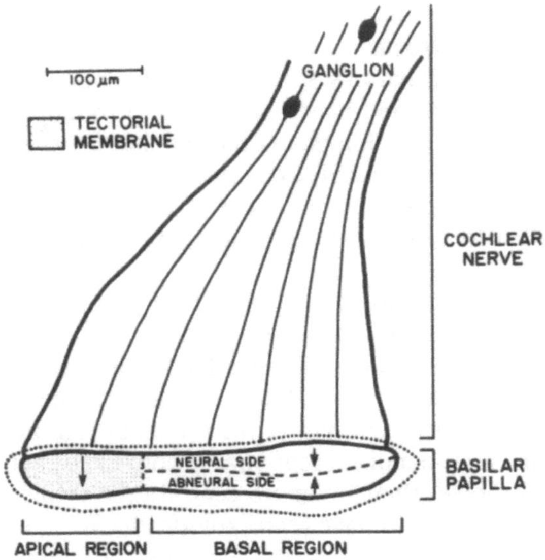

Fig. 11. Entire basilar papilla of the alligator lizard showing location of thick tectorial membrane in apical region and changes in kinociliar polarization from one end to the other (*arrows*). From Weiss et al. (1978)

(Mulroy 1974b) and then at the very basal tip gradually decreasing to about 12 μm. Kinociliar orientation is bidirectional across the papilla, i.e., the kinocilia are on opposite sides of the hair bundles of the sensory cells located on either side of a longitudinal axis throughout the basal portion. Kinocilia are not quite as long as the tallest stereocilia and have a terminal bulb (Fig. 9), which is about twice the size of the stalk (Miller 1973a). Although no measurements are available, the diameters of the stereocilia illustrated by Mulroy (1974b) seem to be 2 to 3 times that of the kinocilium.

Using microelectrode techniques for measuring receptor cell potentials and single unit nerve responses, and the Mössbauer technique for recording basilar membrane movements, Weiss et al. (1974, 1976, 1978) and Holton and Weiss (1978) came to the general conclusion that there was frequency selectivity as well as a tonotopic organization of cochlear nerve fibers but that this was determined by something other than or in addition to the motion of the basilar membrane. The evidence that led them to this conclusion was that the responses to sound derived from single nerve fibers were frequency specific and the tuning curves were sharp, whereas the mechanical tuning curves did not demonstrate either feature. They found that the mechanical tuning curves as determined by the Mössbauer method were more broadly tuned that the neural curves and that the shape of the curve varied little between basal and apical ends irrespective of frequency exposure.

Single nerve fibers had "characteristic frequencies" (CF) and their tuning curves were sharply tuned although there were some differences in tuning curve shape between apical and basal fibers. Fibers coming from the apical region responded to sound frequencies below 1 kHz whereas those coming from the basal region responded to frequencies between 1 and 4 kHz. There were other differences between apical and basal fibers: (1) apical nerve fibers tended to have lower spontaneous discharge rates than basal fibers; and (2) apical fibers exhibited two-tone rate suppression whereas basal fibers did not. Measurements of intracellular hair cell potentials appeared to indicate that they had CF which coincided in position with that of the nerve fibers. From these findings they concluded that there was a distinct dichotomy in the CF of nerve fibers and that this corresponded to the basal and apical regions, but that the higher frequency fibers appeared to have an additional tonotopic organization of their own. They believed that these features were a function of the presence (or absence) of tectorial membrane plus ciliary differences. They also suggested that the special tonotopic high frequency arrangement of the basal nerve fibers could be related to the graded change in ciliary height. These important studies have demonstrated the value of correlating morphological and physiological observations on lower animals where significant structural variations occur.

It is not at all difficult to accept the hypothesis that a tectorial membrane with a limbic attachment is associated with sharply defined neural responses because it has been widely held that the critical feature in hair cell activation in the mammalian organ of Corti is the movement of the hairs by the tectorial membrane. A fibrogelatinous membrane into which the hairs are inserted is an integral part of all other auditory, vestibular, or lateral line receptors. It is surprising, however, that the nontectorial portion of the basilar papilla also produces neural tuning curves with a good degree of definition and that thresholds of nerve fibers from the two regions are not greatly different. It is possible that there are afferent innervation differences between the two regions. Opposing polarization of adjacent cell groups is a feature of the basal portion. If there were a terminal afferent nerve branching pattern which provided for combined excitatory and inhibitory sensory input to a single nerve fiber, sharpening might occur. There do appear to be efferent innervation differences.

Weiss et al. (1978) believe there is a tonotopic organization within the basal region of the papilla and they suggest that the difference in ciliary height which changes from one end to the other might account for this. The thin material that covers the cilia in this region may also be of functional importance. This fibrous covering has been observed by Wever (1978), Miller (1973a), and Mulroy (1974b) and is clearly demonstrated in Miller's illustrations (1973a). The dispersed tectorium plus variations in heights of cilia may be the source of the tonotopic responses, at least in part. If the cilia are completely encased in the gelatinous material (as in the maculae and cristae of the vestibule), then there may be differences in mass from central to basal ends (corresponding to graded changes in ciliary height) which might account for differences in activation. Two other features in the Gerrhonotus ear which may be functionally important are the marked convexity of the papillae and large bulbous kinociliar enlargements which are apparently attached to the stereocilia.

The variabilities presently described are common in reptiles. Kinociliar polarization is quite complex in some lizards (Baird 1974); the tectorial membrane is remarkably varied (Wever 1967a); and vesiculated nerve endings (probably efferent in nature) have been found in restricted areas. Some of these variables are also found in birds and mammals but for the most part reptiles (and especially lizards) seem to have been remarkably versatile in making use of a number of morphological characteristics to derive a maximum of information from a minimum of sensory cells.

4 Birds

The structure and physiological properties of the avian membranous labyrinth have been a source of interest for some years mostly because behavioral and field studies have indicated that birds may have some unusual auditory capabilities. It is obvious that birds are dependent upon sound cues in their daily lives. However, their vocal repertoire is so remarkable that there has been great discussion regarding the quality (frequency, temporal pattern, phrasing, etc.) which is critical in the bird's sound reception. Early histological studies did not clarify how these might be accomplished although they did show that the bird's auditory receptor was somewhat different from that of mammals. The early studies of Retzius (1884) and Held (1926) indicated that the tectorial membrane was massive and that the sensory organ had a cellular organization quite different from that of the organ of Corti in man.

More recently, investigators using ultrastructural methods have given detailed descriptions of the ears of the pigeon (Takasaka and Smith 1971) and the chicken (Tanaka and Smith 1975, 1978). Studies made on other species (Tanaka and Smith, unpublished data) have indicated that the avian ear seems to have developed along more consistent lines than that of reptiles in regard to the specialization of hair cells, innervation pattern, and relation of cilia to tectorial mem-

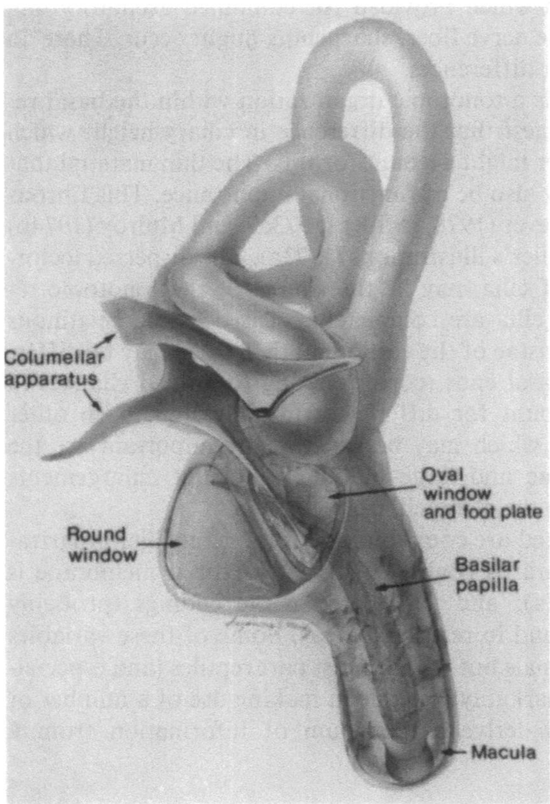

Fig. 12. The chicken's otic capsule and columellar apparatus. The bony capsule of the cochlear duct has been made transparent in the drawing so that the basilar papilla and lagenar macula can be visualized. Semicircular canals are shown above. From Tanaka and Smith (1978)

Table 1. Differences between the small and large birds' papillae in size, measurements of basilar membrane[a], and number of hair cells

	Weight (g)	Length of papilla (mm)	Width of basilar membrane (mm)		No. of hair cells at minimum width	No. of hair cells at maximum width
			Minimum	Maximum		
Pine sisken[b]	12.8	2.2	0.116 (or less)	0.155	11	20
Sparrow[b]	26	2.2	0.06	0.21	3	25
Parakeet[b]	36	2.3	0.07 ± (or less)	0.22	8–9	28
Chicken[c]	1500	4.4	0.07	0.395	3	44
Pigeon[d]	275	3.8			14	54

[a] It was not always possible to accurately measure the minimum width of the basilar membrane; the minimum always occurred at the very proximal tip, which is curved, and those sections were often cut tangential to the long axis of the papilla

[b] Tanaka and Smith (unpublished data)

[c] Tanaka and Smith (1978)

[d] Takasaka and Smith (1971)

brane. Some species variations, however, have been found. A fair amount of physiological and behavioral data have also been produced and the aura of legend that has surrounded birds' vocalizations and hearing is being replaced slowly by fact.

The cochlear duct of the bird is a well-defined tubular sac, even though its proximal end communicates widely with the saccule. The largest proportion of the duct contains the basilar papilla, the distal tip of which is adjacent to a small otolith organ, the macula of the lagena (Fig. 12). Schwartzkopff and Winter (in Schwartzkopff 1968) tabulated the length of the cochlear duct from a number of birds and found that it ranged from approximately 3 mm in small birds to 6–7 mm in larger birds. The owl's cochlear duct is longer than that of most other birds (approximately 12 mm in the barn owl; Konishi and Smith, unpublished

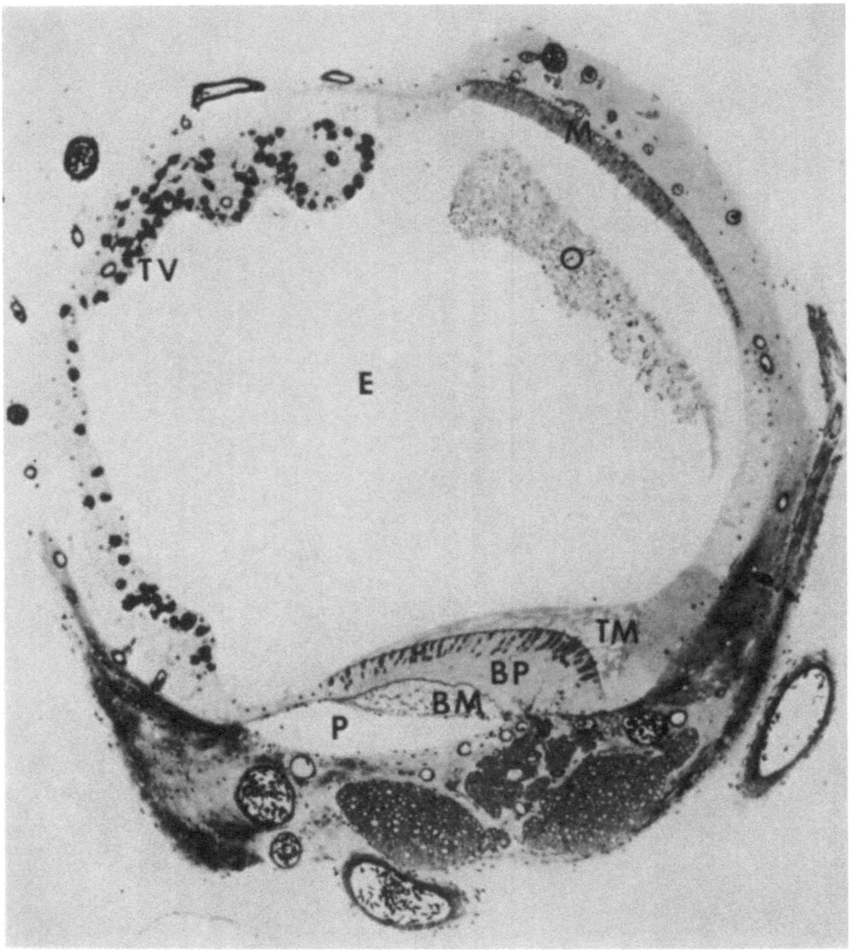

Fig. 13. Cross section of the distal end of the chicken's cochlear duct. The macula (*M*) and otolith membrane (*O*) are shown at *upper right*. The tegmentum vasculosum (*TV*) is at the *left*. The basilar papilla is *below* (*BP*). *BM*, basilar membrane; *E*, endolymph; *P*, perilymph; *TM*, tectorial membrane. ×113. From Tanaka and Smith (1978)

data). The auditory papilla alone is generally about 1 mm shorter. Tanaka and Smith (1978) found that the mean length of seven chicken basilar papillae was 4.3 mm. The lengths of the papillae of several other birds are given in Table 1. It is apparent that the smaller birds (the two passerines and the parakeet) have papillae which are both shorter and narrower than those of the pigeon and the chicken. The basilar membrane of the 12.8-g pine siskin never attains the width nor has as many cells in transverse rows as the 36-g parakeet, but it is doubtful that there is any real correlation with body weight. Head size might be a better comparison,

Even though the avian papillae are short, they contain a large number of sensory cells, many more than would be found on a 4-mm-long mammalian basilar membrane. The pigeon's papilla has approximately 10 000 hair cells (Goodley and Boord 1966) which are arranged in rows with increasing cell numbers ranging from 14 at the proximal end to 54 near the distal tip (Takasaka and Smith 1971). Papillae of the smaller birds probably have about half this number of hair cells (see Table 1).

The papillae of all the birds examined have some common histological features (Figs. 12 – 14). All have a crescentic and spatulate shape. Each is composed of a

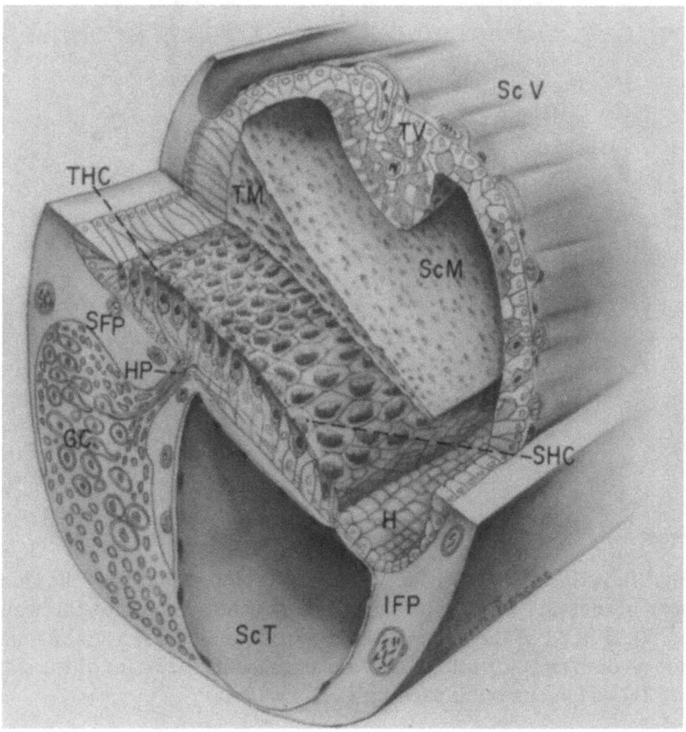

Fig. 14. Section through the central part of the pigeon's cochlear duct. The perilymphatic spaces are scala vestibuli (*Sc V*) and scala tympani (*Sc T*). Scala media (*Sc M*) contains the endolymph; *GC*, ganglion cells; *H*, border cells; *HP*, habenula perforata; *IFP*, inferior fibrous plate; *SFP*, superior fibrous plate; *SHC*, short hair cells; *THC*, tall hair cells; *TM*, tectorial membrane. From Takasaka and Smith (1971)

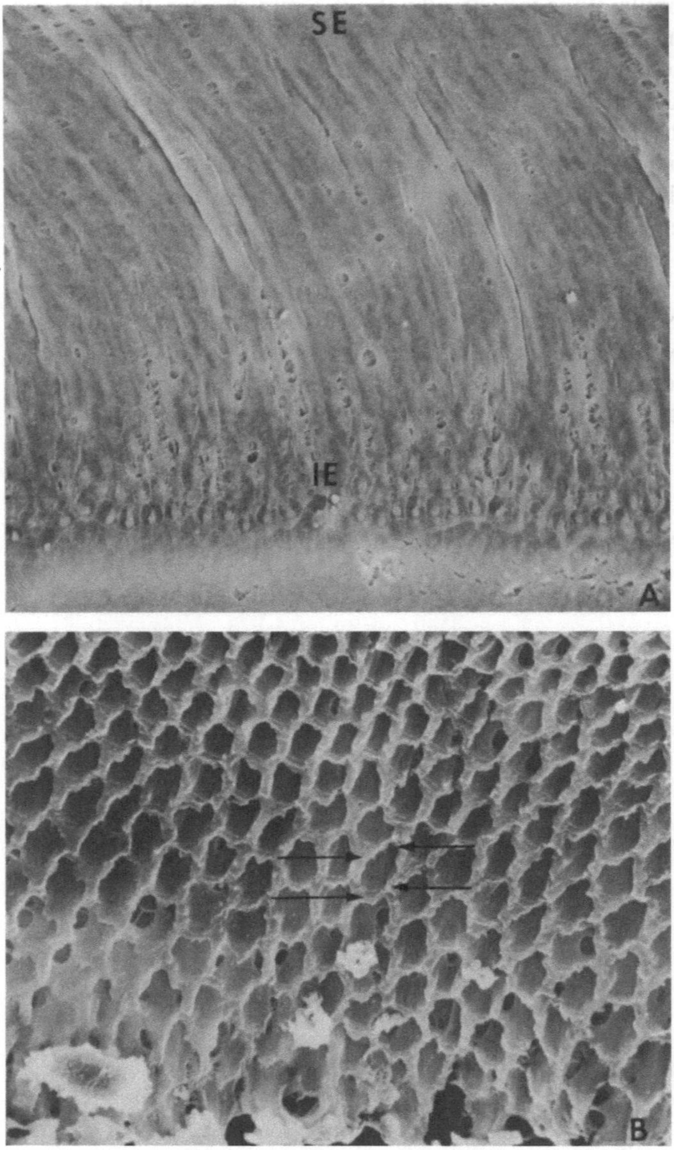

Fig. 15. A SEM of top surface of the tectorial membrane of the chicken. *IE*, inferior edge, near border cells; *SE*, superior edge, at its attachment to the wall of the cochlear duct. ×250. **B** SEM of the under-surface of the chicken's tectorial membrane. The walls of the honeycomb cells (*arrows*) would be attached to the edges of the supporting cells (see Figs. 19, 21) which surround each hair cell. ×1000

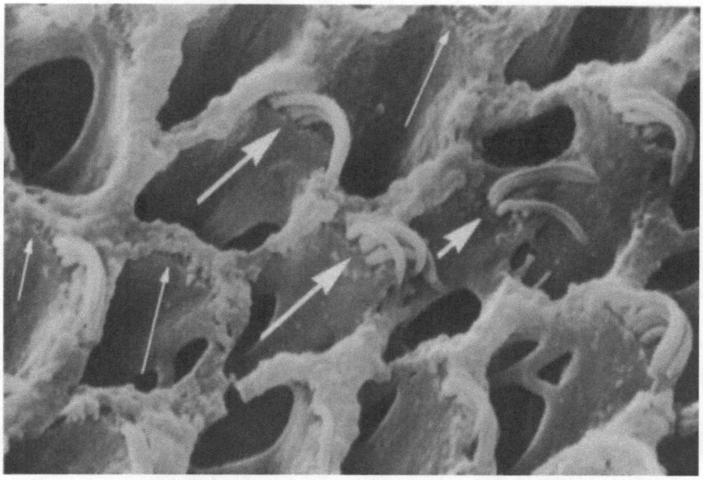

Fig. 16. SEM under-surface of tectorial membrane at higher magnification showing some of the tallest cilia inserted into the membrane (*large arrows*). Note imprints from cilia pulled out of tectorial membrane (*small arrows*). The shorter cilia in the bundle project up into the cavity (unattached to the membrane; see Fig. 19). ×5000. From Tanaka and Smith (1978)

compact mass of sensory and supporting cells. There are several types of hair cells with broadly common locations across species. The basilar membrane changes in width from proximal to distal ends although not always in a regular fashion. All birds seem to have a type of tectorial membrane which is widely attached to the supporting cells, in addition to the hair cell cilia.

The tectorial membrane has a triangular shape in cross section (Figs. 13, 14) with the base of the triangle located at the superior wall of the cochlear duct where it has a broad area of attachment to a group of columnar cells. It covers the papilla and the apex is formed by the attenuated edge of the membrane. Like all tectorial membranes, it is composed of a mass of microfilaments embedded in a dispersed gel. The avian tectorium, however, has some unique features. It is attached to the surface microvilli of all the supporting cells which encircle each hair cell and is generally thicker than the mammalian tectorial membrane (Figs. 15A, B; 16, 19, 21). As a result it presents a honeycomb appearance when viewed from underneath. The hair bundles of the sensory cells project up into the cavities of the honeycomb and the tips of only the tallest cilia on each cell are inserted into the membrane (Figs. 16, 19). The attachment between cilia and tectorium seems to be a strong one because the hairs will sometimes retain their insertion in the tectorium even though they are broken off at the cuticular plate of the hair cells during preparation of the specimens (Fig. 16). The membrane is thinnest at the inferior edge (the apex of the triangle) where it is attached to the epithelial cells at the periphery of the papilla. It is a massive structure in the pigeon (Fig. 14) and chicken whereas in the small birds it is thin and more delicate. But even though it is quite thin at the inferior edge, the avian tectorial membrane resists the stresses of chemical fixation and dehydration and remains firmly in place over the papilla

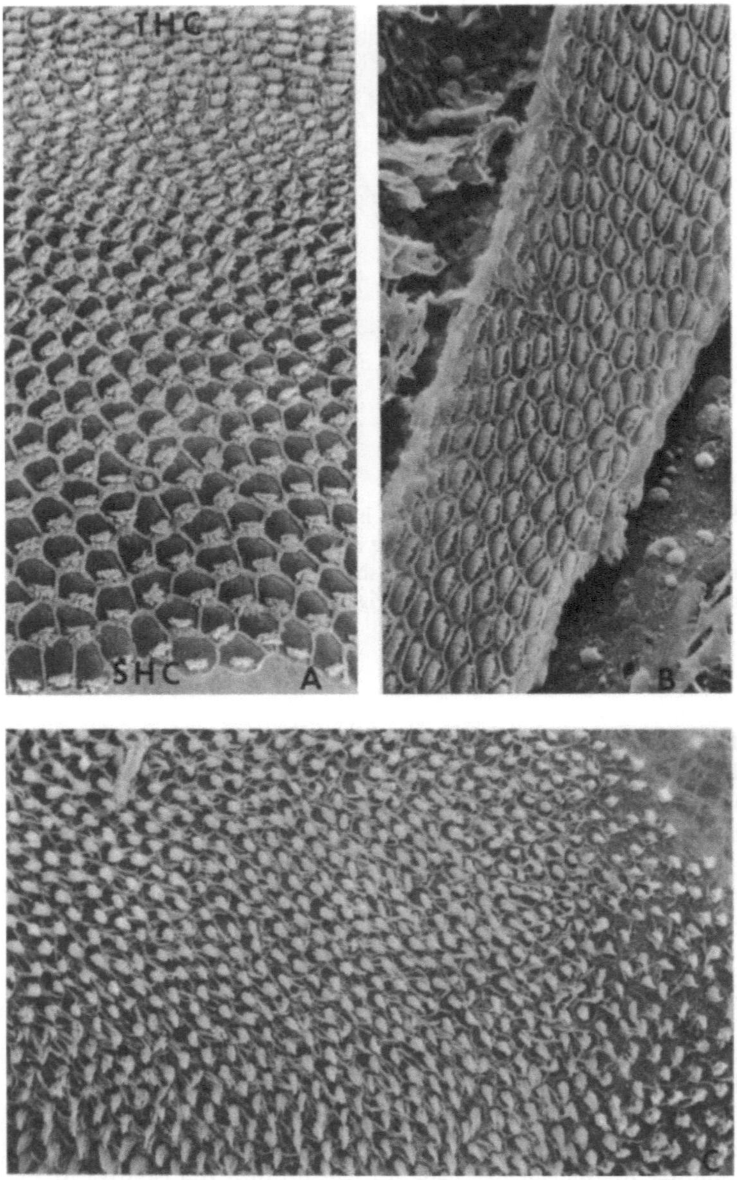

Fig. 17A – C. SEM of surface of chicken basilar papilla. **A** Tall (*THC*) and short hair cells (*SHC*) in central part of papilla. **B** Lenticular hair cells at proximal tip. **C** Tall hair cells at distal tip. **A – C** ×500

(Figs. 13, 15A). This is undoubtedly due to the redundancy of its attachments to the supporting cell microvilli and the tallest sensory cell cilia.

Takasaka and Smith (1971) demonstrated that there were three types of hair cells in the pigeon (Fig. 14) and that these had a specific location on the basilar membrane. This was confirmed in the chicken ear (Tanaka and Smith 1978) where an

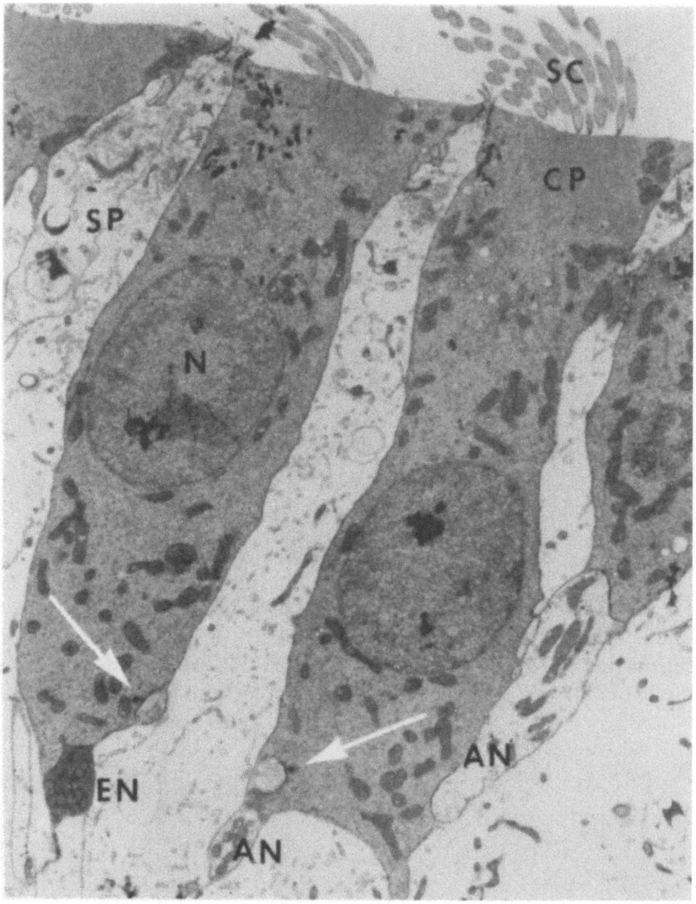

Fig. 18. Transmission electron micrograph (*TEM*) of tall hair cells from the chicken basilar papilla. Synaptic ball structures (*arrows*) adjacent to afferent nerve endings (*AN*). *CP*, cuticular plate; *EN*, efferent nerve ending; *N*, nucleus; *SP*, supporting cell; *SC*, stereocilia. ×5000. From Tanaka and Smith (1978)

additional type (the lenticular cell) was found (Figs. 17A – C). The tall hair cells are columnar in shape (Fig. 18). They are found at the superior edge of the papilla and also completely cover the distal tip in the pigeon and chicken. They have a number of afferent nerve endings, some quite large, and of variable configuration. Only on occasional small efferent bouton is present. On the other hand, the short hair cells have very large efferent nerve endings (Fig. 19) which form such a huge basal chalice that the only spaces left for the small afferent nerve endings are on the sides of the hair cells. The short hair cells are pitcher shaped with large apical surfaces and are located on the portion of the basilar membrane suspended over the perilymph. The type of cell at the distal tip varies between species.

A third type is the intermediate cell. This is a transition type of cell and is usually located between tall and short hair cells. It is intermediate in height without the

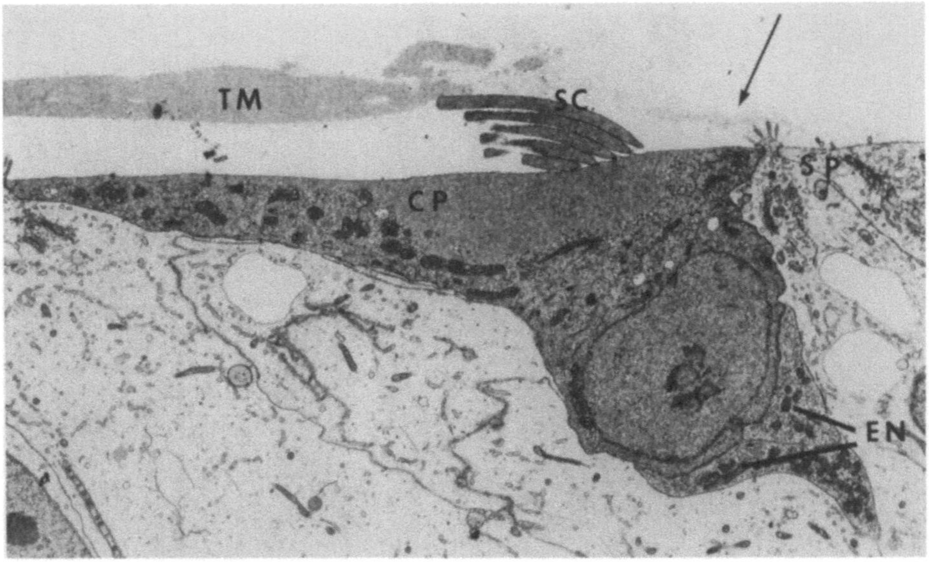

Fig. 19. TEM of short hair cell from chicken basilar papilla with two large efferent endings (*EN*). No afferent nerve endings are present in this section. Note the microvilli tufts on supporting cell (*SP*) at right with attached fibrous extensions (*arrow*) from tectorial membrane (*TM*); *CP*, cuticular plate; *SC*, stereocilia. ×5000. From Tanaka and Smith (1978)

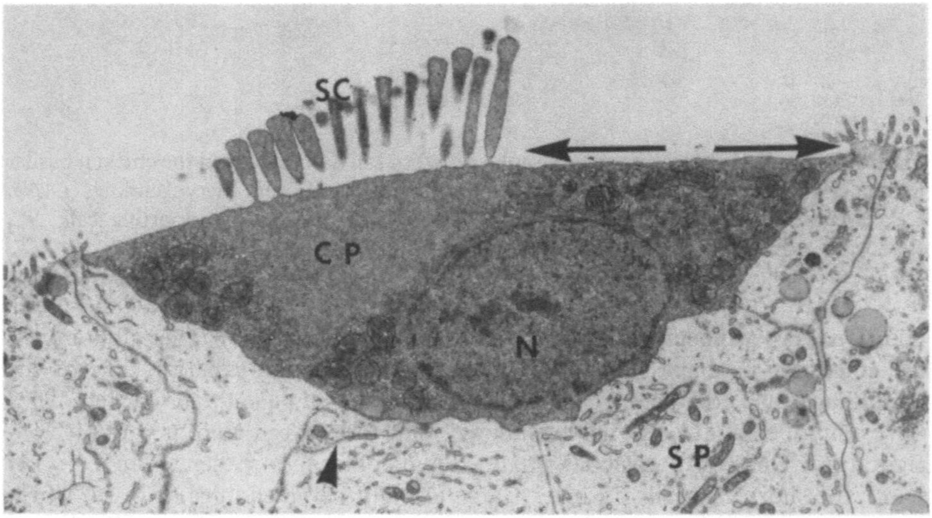

Fig. 20. TEM of lenticular hair cell from chicken's basilar papilla. Note large cuticular-free area at right (*double arrows*). *Arrow head,* small afferent nerve ending; *N*, nucleus; *SP*, supporting cell; *SC*, stereocilia. ×10300. From Tanaka and Smith (1978)

enlarged apical surface and usually has moderate-sized afferent and efferent nerve endings.

A fourth type of cell, the lenticular cell, was found by Tanaka and Smith (1978) at the extreme proximal end of the chicken papilla (Figs. 17C, 20). This sensory cell is distinguished by an "island" cuticular plate. The apical end of the cell is broad and flat but the fibrous cuticular plate is either off to one side or only centrally located, and it covers no more than about one-half of the cell's apical surface.

The number of cilia per bundle and the lengths of the cilia are not equivalent in all the cells. Tanaka and Smith (1978) observed that, in the chicken, numerical variations seemed to be related to location on the papilla rather than to type of cell. For example, the largest number of cilia per bundle were present at the proximal end (160 – 170) and this was where the lenticular cells were located. These cilia were short (2.5 μm tall) and stubby. The smallest number (50 – 65) was on the tall cells at the distal end and these were also the longest (8 μm). In the central part, there was not a remarkable difference between numbers of cilia on different cell types (90 – 120 on tall and 80 – 105 on short cells).

The ciliary bundle contains a single kinocilium which is located adjacent to the central notch in the tallest row of stereocilia (Fig. 21) with its basal body in a small cuticular-free patch. Takasaka and Smith (1971) found that every hair bundle in the pigeon papilla had a kinocilium and that the direction of polarization was fairly consistent throughout the papilla, i.e., toward the inferior edge. This unipolar arrangement was also present in the chicken (Tanaka and Smith

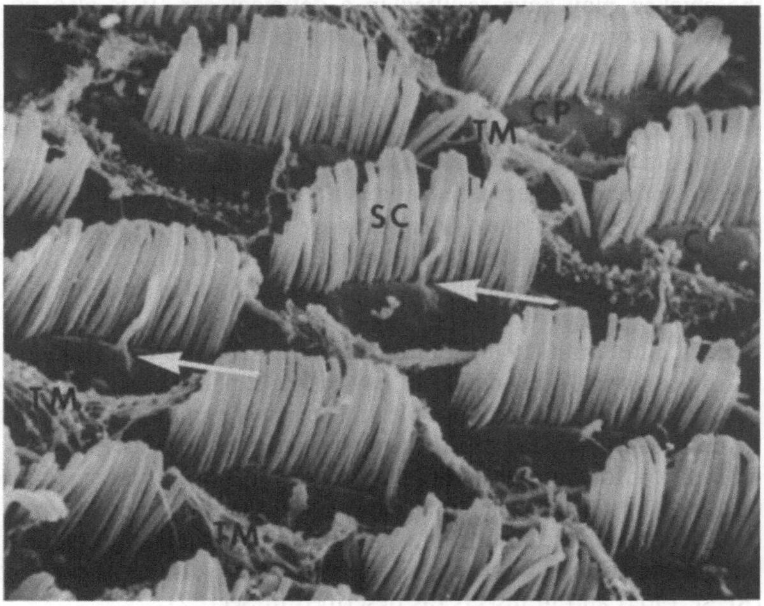

Fig. 21. SEM of surface of chicken's basilar papilla showing kinocilia on two hair cells in center (*arrows*). Other hair cells show only stereocilia (*SC*). Note remnants of tectorial membrane (*TM*) attached to microvilli of supporting cells, which encircle each hair cell. *CP*, cuticular plate. ×5000. From Tanaka and Smith (1978)

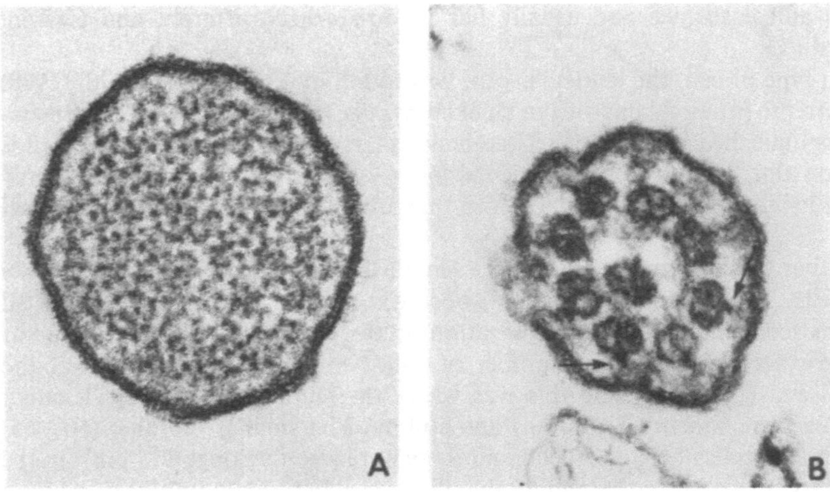

Fig. 22A, B. TEM showing cross sections of A stereocilium with closely packed microfilaments and **B** a kinocilium with eight peripheral and one central doublet. *Arrow* at small projections, probably dynien arms. **A** and **B**, ×204000. From Tanaka and Smith (1978)

1978), but many hair bundles in the chicken had no kinocilia at all (Fig. 21). Information on the incidence of kinocilia in other birds is not available at present. It is possible that the breeding processes which have followed domestication of the chicken may have resulted in a deficient ear. It is also possible that the kinocilia play no active part in effective operation of the avian auditory receptor. The mammalian cochlea performs very well without kinocilia.

There has been some controversy about the internal structure of the auditory and vestibular kinocilia. (Each vestibular hair bundle in the vertebrate ear does have a kinocilium.) Flock (1964) and Flock and Wersäll (1962) have clearly shown that the kinocilia in the lateral line and in the vestibule of the fish (*Lota vulgaris*) have a 9 + 2 structure. Some dynien arms are visible in their micrographs but it is not clear if these are of the usual configuration. Hamilton (1969) has shown that kinocilia in the rat vestibule have irregularities and are not identical with respiratory cilia.

Tanaka and Smith (1978) found that the chicken's auditory kinocilia were irregular in construction and did not have a 9 + 2 pattern (Fig. 22B). Both peripheral and central tubules were doublets. There were small dynien-like arms but they tended to project directly toward the peripheral membrane and to be derived from a position between subfibers A and B, instead of from subfiber A. The pattern most frequently encountered was 8 + 1 although 6 + 1, 6 + 2, and 7 + 1 combinations were also seen. These variations are quite similar to kinociliar adaptations seen in other sensory structures such as the olfactory receptors (Reese 1965) and in insects (as described in Sect. 2).

The width of the basilar membrane as well as the number of sensory cells in the pigeon (Takasaka and Smith 1971) and chicken (Tanaka and Smith 1978) increases from proximal to distal ends. In addition, Tanaka and Smith (1978) observed that the thickness of the chicken basilar membrane also increased from a thin membrane covered by a flattened sensory organ at the proximal end to a thickened membrane overlaid with a thick, convex sensory structure at the distal end. These features plus an increasing number of cells seem to be part of a regular, changing pattern along the basilar membrane from proximal to distal ends of the papilla that fits in well with the salient requisites for the propagation of a traveling wave. Bekesy (1960) made measurements on the chicken's ear and found that the place of vibration along the basilar membrane moved from apical to basal end with frequency exposures of 0.1 kHz to 3 kHz. He believed that no frequency analysis was possible below 0.1 kHz because the cochlear portion vibrated as a whole at this frequency and below.

There is no evidence as to whether a traveling wave occurs on the 2-mm papillae found in smaller birds. The change of basilar membrane width from proximal to distal tip is approximately a two- or threefold one (Table 1), but this does not occur gradually. For example, measurements (Smith and Tanaka, unpublished data) made from sections of a parakeet papilla revealed that, about halfway along the papilla, the basilar membrane was already 0.19 mm wide and increased only slightly more to 0.22 mm. Just before rounding off at the distal tip it decreased to 0.195 mm. The width (and cellular mass) therefore changed very little in the distal one-half of the parakeet's basilar membrane.

It is possible that frequency discrimination is accomplished by some means other than differential basilar membrane movements or that frequency discrimination except over a limited range is really not important to some birds. If one compares the cat ear where 12500 cells are spread out over 23 mm of basilar membrane, or the rabbit, whose 9000 cells occupy 15 mm (Retzius 1884), with the pigeon's 10000 cells on 4 mm, obviously something quite different is occurring. Perhaps innervation differences, arrangement of sensory cells, or other features are more important in the very short avian papillae.

Indeed, recent studies on the pine sisken, sparrow, and parakeet have indicated that there are variations in the distribution of the various cell types (Smith and Tanaka 1981). The three major types of hair cells (tall, intermediate, and short) were readily identified in each of these birds: the tall hair cells occupied a position close to the attachment of the tectorial membrane whereas the intermediate and short hair cells were located over the free basilar membrane in the central part of each papilla. At the distal tips the cell type differed between species. A large population of tall cells covered the distal tip in the pigeon. This region in the pine siskin was filled with intermediate cells. The sparrow had a few short hair cells at the tip but the parakeet had a large number. The continuation of three types of cells in transverse rows across the basilar membrane was a pattern that had a much more distal extension in the parakeet than in any of the other species examined. Clearly more detailed studies are necessary on other individuals and other species. Basilar membrane dimensions and locations of cell types may be important features in the differences between species in frequency reception. Details of innervation such as peripheral branching patterns and central projections are other essential pieces of information as yet available.

Within the past 10 years an accumulating body of evidence has indicated that audition for most birds is best over a limited frequency range. Konishi (1970) measured the neural responses to sound in the brainstem nuclei (nucleus magnocellularis and nucleus angularis) of a number of birds and found that the largest number of units sampled (which were also those with the lowest thresholds) responded to a limited frequency range which was generally no more than two octaves. It varied between birds and, to give two extreme examples, was between 1 – 3 kHz for the house sparrow and 2 – 9 kHz for the slate-colored junco. Konishi hypothesized that hearing in birds was related mostly to species recognition and found that their best frequency sensitivity correlated fairly well with the frequency spectrum of their calls and/or songs.

Sachs et al. (1974) studied the unit responses from the cochlear nerve of the pigeon and reported a somewhat broader band: 0.25 – 4 kHz. Although some of the most sensitive units were clustered around 2 kHz, a far larger number of nerve fibers were sampled which responded to frequencies below 2 kHz. This could well reflect the large number of hair cells located on the distal two-thirds of the pigeon's papilla (Takasaka and Smith 1971). Cochlear nerve fibers are more numerous at the distal end, also.

Audibility curves obtained by behavioral methods are available for a number of birds, and most show the same type of limited frequency range as found by neurophysiological monitors (Fig. 23). Stebbins (1970) found that two pigeons gave responses that peaked at 1 kHz. Dooling and his associates (1971, 1975, 1978) have trained birds of several different species by identical methods and found the parakeet to be more sensitive than the canary or the house finch. All three birds had a narrow range which fell off sharply above 5 – 6 kHz and peaked at about 2 – 3 kHz. The trend, as presently available, points toward a low frequency avian receptor organ with a sharp sensitivity peak and a limited range. One notable exception to this is the barn owl (Fig. 23), which has a long papilla and a frequency range of several octaves (Konishi 1973). It may be that, as more birds are examined, other exceptions will be found.

One major difference that Sachs et al. (1974) found between neural responses from pigeon and cat was the higher spontaneous activity displayed by the pigeon. If the spontaneous activity is related to spontaneous release of transmitter substance the higher activity in pigeons could be explained by the large synaptic balls and a different innervation pattern. It is known that cat inner hair cells are innervated by about 90% of the cochlear nerve fibers (Spoendlin 1974) and that the

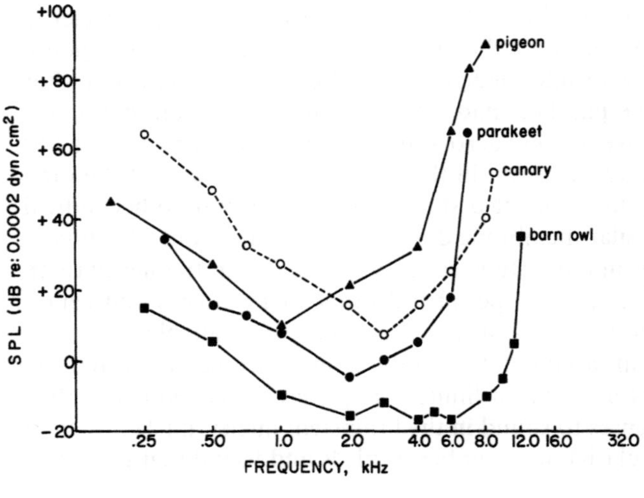

Fig. 23. Audibility curves from four birds. The barn owl's sensitivity is very good over a broader range than that of other birds. The pigeon's curve is from Stebbins 1970; the parakeet's is from Dooling 1973; the canary's is from Dooling et al. 1971; the barn owl's is from Konishi 1973

inner hair cell afferents in the cat rarely supply more than one or two sensory cells (Spoendlin 1974; Liberman 1979). At present we have no detailed data on terminal branching patterns of afferent nerve fibers in the pigeon but other data are available. There are a large number of afferent fibers at the distal end of the pigeon papilla which is populated by tall hair cells. The ratio of 5000 cochlear nerve fibers to 10000 hair cells in the pigeon (Boord 1964; Goodley and Boord 1966) is quite different from the cat with 50000 nerve fibers to 12000 hair cells. It seems likely that in the pigeon a single nerve fiber is connected to more hair cells than in the cat. In addition, the nerve endings are large, sometimes form a shallow cup about the base of the hair cell, and in the chicken, they often invaginate into the hair cell (Tanaka and Smith 1978). There are large synaptic ball structures, sometimes two or three adjacent to a single terminal in both pigeon and chicken hair cells. This large synaptic area and more synaptic bodies could increase possibilities for release of transmitter substances and result in higher spontaneous activity rates.

In some ways, the vestibular synapses are rather similar to those on the tall hair cells of the avian auditory receptors. Sensory cells in both organs have large nerve endings with invaginating processes and with many synaptic bar or ball structures. Goldberg and Fernandez (1971) found that the resting discharge rate of the vestibular nerve fibers was also high. Although it may not be valid to compare data obtained under different conditions, the discharge rates for squirrel monkey cristae reported by Goldberg and Fernandez are almost identical to the rate level obtained by Sachs et al. for the pigeon (1974).

5 Mammals

5.1 Organ of Corti

The organ of Corti of the mammalian ear differs from the auditory receptor organs of other animals in that its structure is remarkably consistent from one end of the cochlea to the other. There is a single row of inner hair cells medial to a fluid-filled tunnel whose walls are formed by the tunnel rod cells. The outer hair cells are located lateral to the tunnel and are arranged in three rows. Occasionally there is a fourth hair cell but not a fourth row. The organ of Corti is similar to reptilian, avian, and even insect receptors in its stiffness that is due in great part to the presence of microtubules.

The sensory cells and their supporting cells which form the organ of Corti are situated on a basilar membrane as in other vertebrates, but the mammalian structure is coiled around a central core of bone called the modiolus. This houses the nerve fibers and blood vessels. Even the short 7-mm basilar membrane of the mouse (Bekesy 1960) is coiled in this fashion. Only the mammalian auditory apparatus has a true cochleate form and deserves to be called a cochlea.

The length of the basilar membrane and the number of hair cells varies with species. In man the basilar membrane is 34 mm long and there are approximately 16800 hair cells. The 3400 inner hair cells are arranged in a single row medial to the tunnel and the 13400 outer hair cells in three rows lateral to the tunnel (Bred-

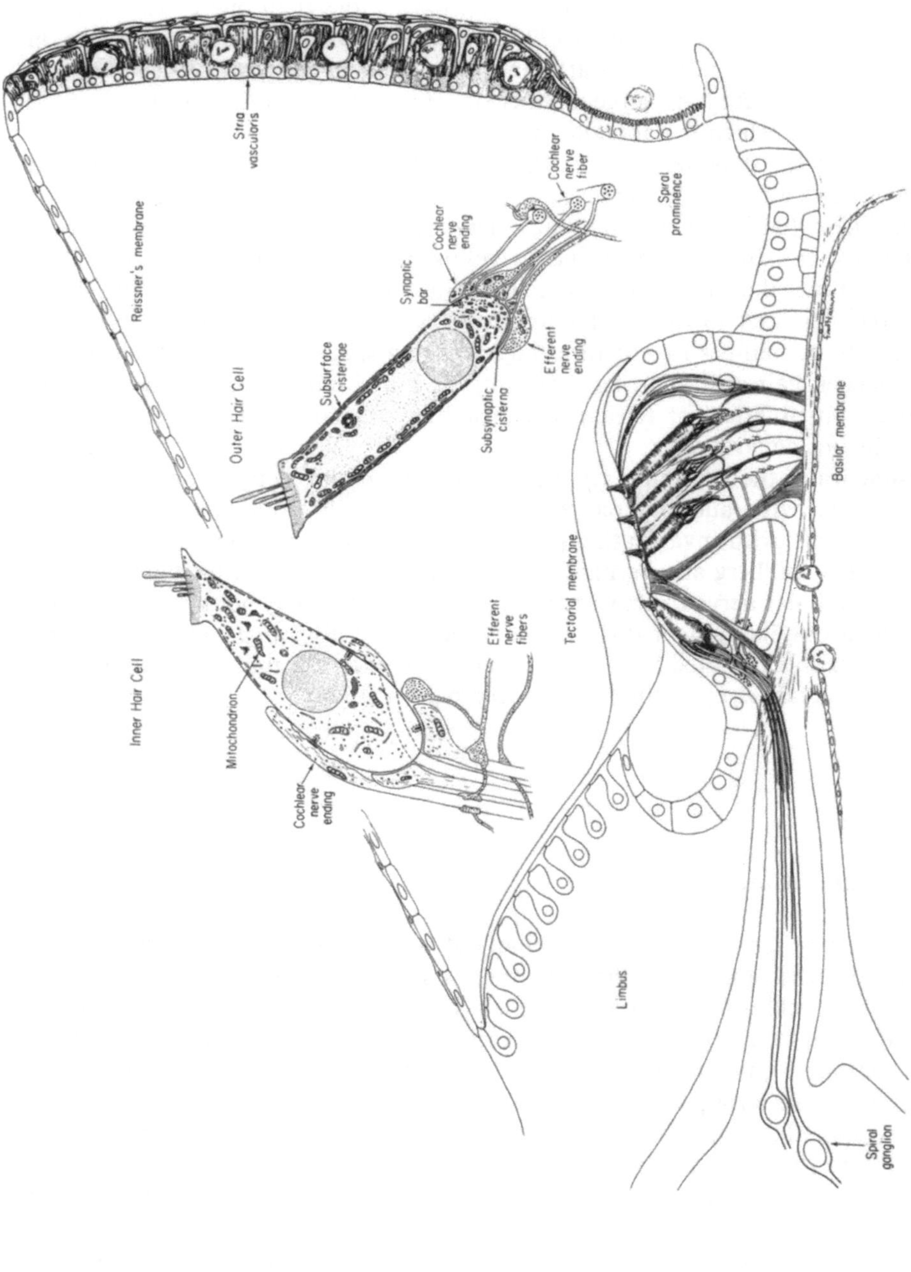

Fig. 24. Cross section through the cochlear duct of the guinea pig. Drawing by Fred Harwin. From Smith (1975b)

Fig. 25. SEM of the guinea pig cochlea. The bone of the otic capsule has been removed. The stria vascularis, Reissner's membrane, and the tectorial membrane have been dissected away to expose the organ of Corti (*arrowheads*). Some remnants of the medial insertion of Reissner's membrane (*RM*) are still attached to the bony wall of the modiolus (*MO*). Area in ellipse (*at tip of white arrow*) enlarged in Fig. 26. ×20. From Smith (to be published)

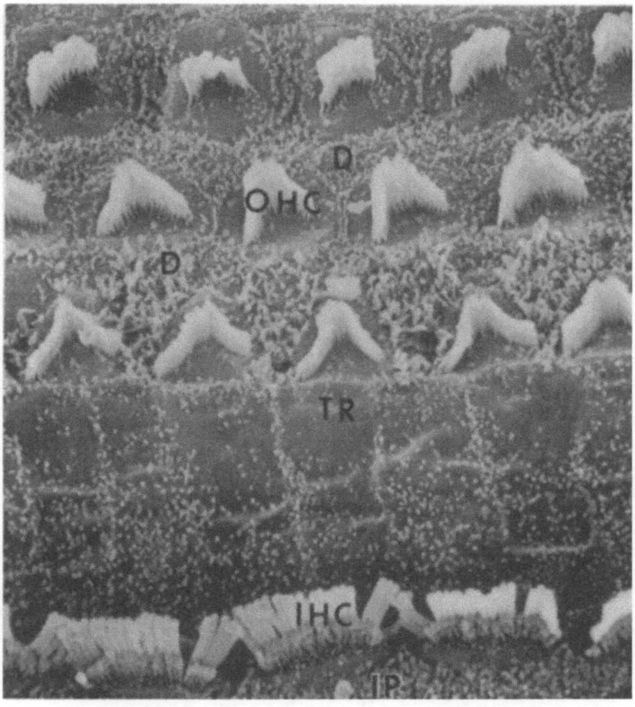

Fig. 26. Organ of Corti. Higher magnification of area designated in Fig. 25. The hair bundles on inner (*IHC*) and outer (*OHC*) hair cells are obvious. Smaller protrusions are the microvilli on the inner phalangeal cells at extreme lower right (*IP*), the tunnel rod cells (*TR*), and Deiters cells (*D*). ×3000. From Smith (to be published)

berg 1968). Irrelevant of number, the hair cells are always arranged in a single row of inner and three rows of outer hair cells throughout the cochlea (Figs. 24–26).

5.1.1 Inner Hair Cells

Cell Structure. The use of newer microscopic techniques has added some details to our knowledge of the morphology of the sensory cells in the past decade, but most of the significant features were revealed in earlier electron microscopic studies.

The inner hair cell is a flask-shaped cell (Figs. 3, 24, 27). The apical end is somewhat oval in form and the hairs (or cilia) are arranged in a few straight rows which almost bisect the long axis of the cuticular plate (Fig. 26). The longest hairs on the inner hair cell measure from 2.3 μm (basal coil of the cochlea) to 4.3 μm (apical coil) in the rat (Iurato 1967). The diameter of the inner hair cell cilia is about twice that of cilia on the outer hair cells (Iurato 1967). Furthermore, the

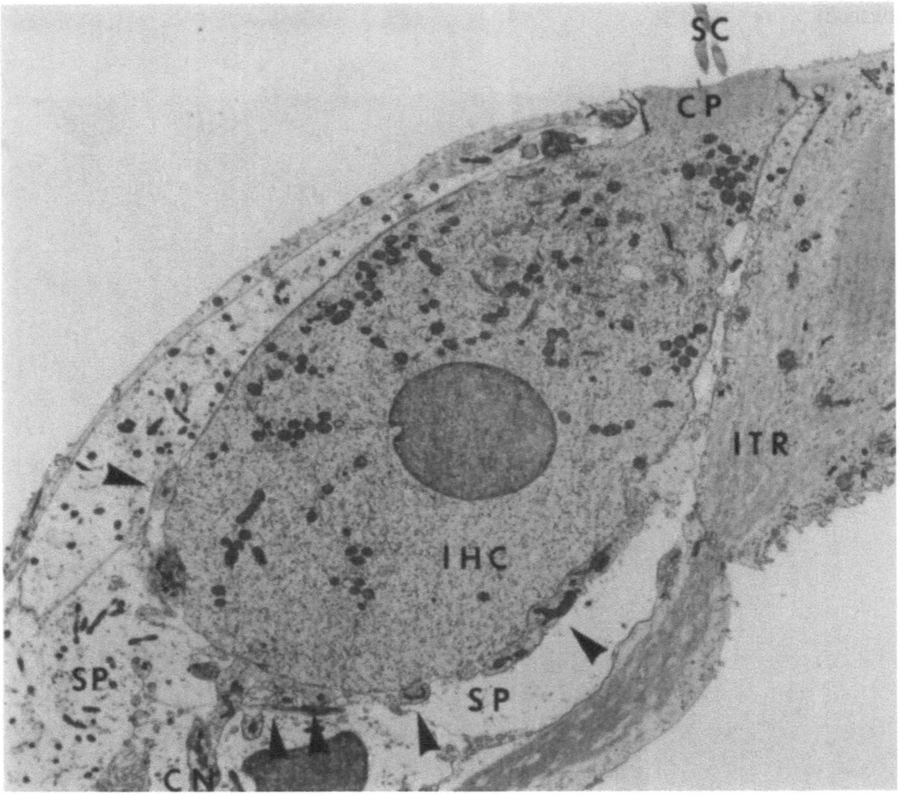

Fig. 27. TEM of inner hair cell (*IHC*) of the gerbil. Unmyelinated cochlear nerve fibers (*CN*) in between the supporting cells (*SP*) at *lower left*. Several nerve endings are visible around the basal end (*arrowheads*); one long fingerlike ending at *right* (*arrowhead*). *CP*, cuticular plate; *ITR*, inner tunnel rod cell; *SC*, stereocilia. ×4300. From Smith (1978)

former are flattened on top. All the hairs are stereocilia. A single kinocilium is present just outside the cuticular plate in the fetus (Kikuchi and Hilding 1965) but this disappears at maturation. Some basal bodies are retained but these have been seen so infrequently that it is questionable that they have any real significance. A single row of small vesicles lines the plasma membrane of the apical half of the hair cell. There is a large amount of Golgi apparatus but otherwise the cytoplasmic components of the inner sensory cells are not remarkably different in quantity or arrangement from those of other cells. What is remarkable is that the inner hair cell and its nerve endings may well be the most consistent feature in the mammalian organ of Corti. Inner hair cells show little, if any, modification across species.

Innervation. The nerve endings are found around the enlarged basal half of the cell. Innervation characteristics of the guinea pig inner hair cells were clarified (Smith 1961) by means of serial sections. The terminals take different forms which include boutons, large clublike terminals, and long fingerlike processes up to 9 μm long which ascend the sides of the hair cells. Synaptic bar structures are prevalent within the hair cell cytoplasm adjacent to presynaptic membrane densities (Smith and Sjöstrand 1961a).

A single fingerlike terminal is not in close contact with the hair cell for its entire length. Rather, membrane thickenings, close apposition, and presynaptic bars give evidence for synaptic activity only at restricted regions. It is possible that these synaptic patches are dynamic features and do not maintain a constant location along the membrane. This would imply that the synaptic bar is not attached permanently to the presynaptic membrane. It is also possible that all the patches on a single nerve fiber are not simultaneously active.

Liberman (1979, 1980) has recently made serial sections of the cat's inner hair cells and found similar types of nerve endings. He found that most radial fibers in the cat had synaptic thickenings only at their terminal bulbous tips and that synaptic bars were adjacent to the membrane densities, within the hair cells. He had previously (1978) measured single unit responses from the cochlear nerve of the cat and observed that spontaneous activity (SA) fell into three groups: low, intermediate, and high. His studies on fiber morphology (1980) indicated that the unmyelinated preterminals to any one inner hair cell could be classified by size and density of mitochondria. He suggested that the mitochondria-rich fibers might be related to the high SA fiber responses previously recorded (1978). One would be somewhat more comfortable with a morphological explanation that included synaptic variables. He is in agreement with Spoendlin (1970) that the radial nerve fibers branch little in the cat. However, we have seen many examples of single long nerve ending with synapses on two neighboring inner hair cells in the chinchilla, a feature which could have the same functional significance as bifurcation.

Some efferent nerve fibers do terminate on the inner hair cells but these are rare, at least in the guinea pig (Smith 1975a). The inner hair cell efferents course in the inner spiral bundle which is composed mostly of ipsilateral olivocochlear fibers and has multiple axodendritic synapses on the afferent cochlear nerve fibers just below the inner hair cells.

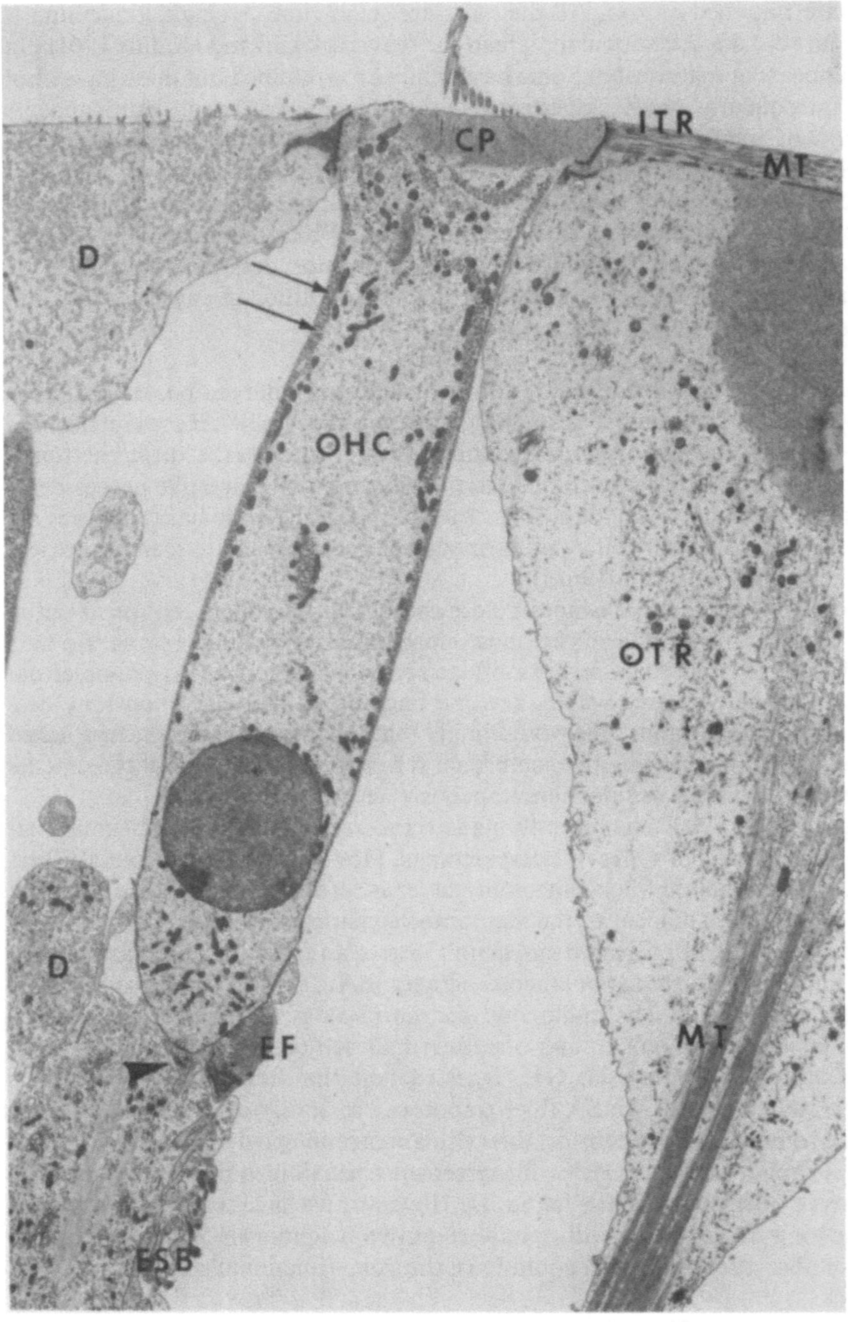

Fig. 28. TEM of outer hair cell (*OHC*), first row, guinea pig. Note the bundles of microtubules (*MT*) in the outer tunnel rod (*OTR*) as well as in the extension of the inner tunnel rod (*ITR*) which joins the outer hair cell. Multiple subsurface cisternae at *arrows*; cochlear nerve ending at *arrowhead*. *CP*, cuticular plate; *D*, Deiters' cells; *ESB*, unmyelinated cochlear nerve fibers in external spiral bundle; *EF*, efferent nerve ending. ×4300. From Smith (1978)

5.1.2 Outer Hair Cells

Cell Structure. The outer hair cells in the mammalian cochlea (Figs. 3, 24, 28) are the most highly differentiated of all the auditory receptor cells. They have the characteristic features of cuticular plate and cilia as do the other hair cells but there are membrane specializations (internal and external) not present in the others. Furthermore, organization of the cytoplasmic elements is different above and below the nucleus.

The cells are tall and columnar in form with the nucleus located always in the basal part of the cell. In the supranuclear region, the mitochondria and the endoplasmic reticulum are found mainly at the cell periphery, i.e., just beneath the cuticular plate and close to the plasma membrane. The apical membranes take the form of smooth vesicles and cisternae or Golgi apparatus. The membranes at the sides of the cell form one or more layers (depending upon species) of cisternae which are adjacent to the plasma membrane and connected to it by minute electron-dense bridges (Smith 1978). Mitochondria are located adjacent to the innermost of these subsurface cisternae. Very few mitochondria or other organelles are present within the central part of the cytoplasm. The plasma membrane is also unusual in that it is serrated, thick, and less electron dense than other plasma membranes (Smith and Dempsey 1957). Recently Gulley and Reese (1977) have found that the particulate nature of this plasma membrane as revealed by the freeze-fracture technique is quantitatively different from that of the inner hair cell.

The infranuclear part of the outer hair cell is filled with membranes and mitochondria. The cytoplasmic volume and, consequently, the quantity of organelles vary between species. The infranuclear cytoplasmic area in the gerbil, for example, is so small that the few mitochondria present indent the nuclear membrane (Smith 1978).

Innervation. The basal part of the cell is encased in a shallow nerve chalice which is composed of closely packed afferent and efferent nerve endings (Fig. 29). The *efferent* nerve terminals on outer cells in all animals examined are large and most numerous in the basal coil of the cochlea (Smith 1973). The number per hair cell varies among species: There are more than ten of the large (0.7 – 3 μm diameter at the synapse) efferent endings on the majority of outer hair cells in the guinea pig's cochlea (Smith and Sjöstrand 1961b) but only one to three per cell in the monkey (Smith 1973) and the chinchilla (Iurato et al. 1978). The efferent terminal bulbs are packed with mitochondria and small vesicles.

The number of *afferent* nerve endings likewise varies with location in the guinea pig ear. More than 30 per cell were counted in an incomplete series of sections from the apical part of the cochlea; there are considerably fewer afferent terminals per cell in the basal coil (Smith and Sjöstrand 1961b). However, the larger number of nerve endings does not necessarily mean there are a larger number of cochlear nerve fibers to the apical coils. It is probably a reflection of the larger number of *terminals* per nerve fiber in the apex (Smith 1975a). Morrison et al. (1975) have given evidence that the number of afferent fibers to all parts of the guinea pig cochlea is fairly constant. Kimura et al. (1964) found that the number of nerve endings of any kind was low on the outer hair cells in man.

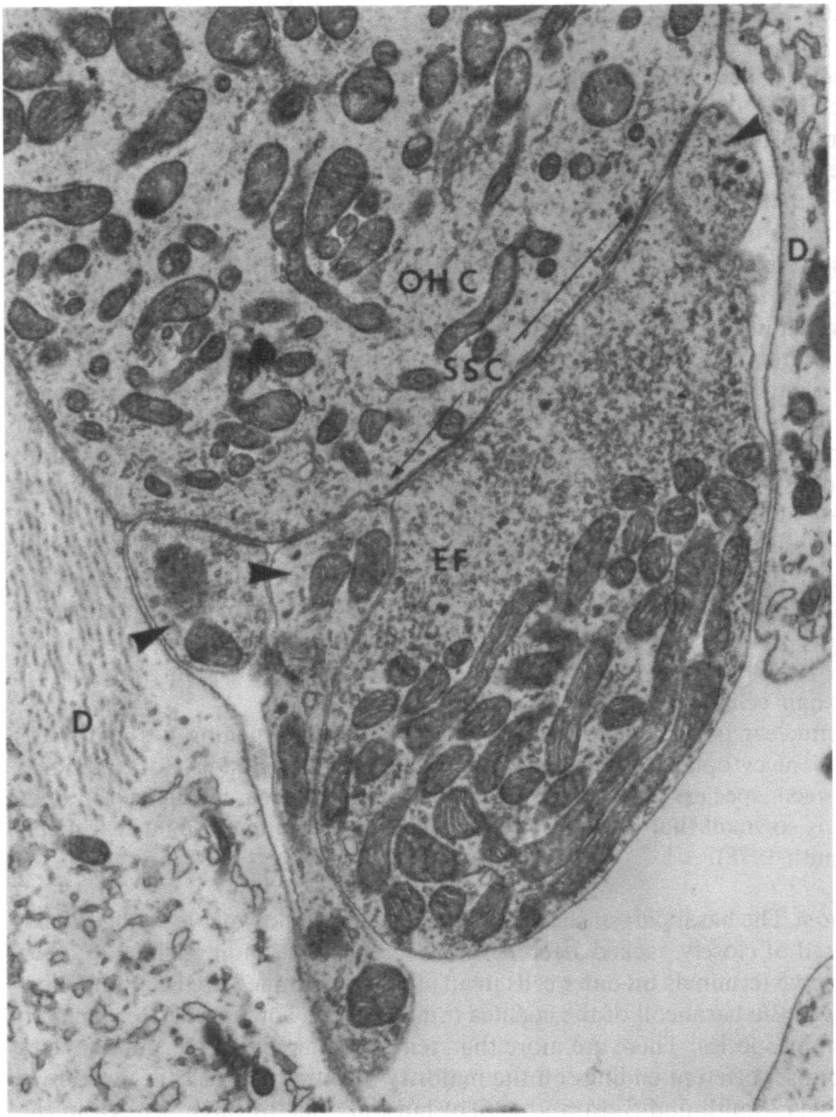

Fig. 29. Nerve ending on the basal end of outer hair cell (*OHC*), chinchilla, with three small cochlear nerve endings (*arrowheads*) and one large efferent nerve ending (*EF*). *D*, Deiters' cell; *SSC*, subsynaptic cisterna. ×24800. From Iurato et al. (1978)

Certain features are characteristic of the outer hair cell synapses. A synaptic bar structure is generally present within the hair cell cytoplasm opposite each afferent terminal in the guinea pig (Smith and Sjöstrand 1961a, b). Dunn and Morest (1975) have reported that synaptic bars are not present in the cat outer hair cells whereas they are readily visualized within the inner hair cells. Small- and medium-sized vesicles are present in the basal hair cell cytoplasm and in the sensory endings, but the vesicles are scattered and not consistently near the synapse.

A large flat subsynaptic cisterna without attached ribosomes is always present adjacent to the postsynaptic membrane of the efferent terminals. This cisterna is apparently attached to the inner leaflet of the hair cell plasma membrane because it remains in place even several months after the efferent terminals have degenerated following transection of the nerve trunk (Smith and Rasmussen 1963; Iurato et al. 1978). The chinchilla (Smith and Rasmussen 1963; Iurato et al. 1978) has many axodendritic synapses below the outer hair cell region. Kimura and Schuknecht (1964) found these also in man. Spoendlin (1970) has not been able to find axodendritic synapses in this region in the cat.

5.1.3 Neural Patterns

Afferent. The spiral ganglion is housed in a bony canal which winds around the modiolus from base to apex, and its constituent cells are the first-order neurons of the cochlear nerve. The neurons are completely myelinated in most mammals, although the thickness of the myelin sheath varies (Rosenbluth 1967; Spoendlin 1974). There could be several important functional correlates of a myelinated bipolar neuron. One obvious one would be faster conduction from cochlea to cochlear nucleus. Another would be the improbability of any somatic synapses. And if the myelin sheath were continuous, dendritic and axon synapses would also be lacking. Dendritic synapses would be prevented near the cell body of the cochlear neurons but would be possible along the terminal *unsheathed* portion within the organ of Corti. Indeed, there are numerous synaptic plaques between olivocochlear axons and inner hair cell afferents beneath the inner hair cells and in some animals under the outer hair cells. It has not yet been documented that any synapses exist centrally from the habenula perforata to the terminal arborizations of the cochlear nerve axons at the cochlear nucleus in lower mammals.

Kimura et al. (1979) have found recently that many neurons in the spiral ganglion of man are unmyelinated. They have detected synapses on some of the small ganglion cells but the identity of the latter cells is uncertain at present.

Lorente de No's original Golgi studies on the rat cochlea (1937) showed that the peripheral processes of the spiral ganglion cells were distributed either to the inner (radial fibers) or to the outer hair cells (the external spiral fibers) and that two distinct and separate neural patterns were thus formed. More recently the question that a single spiral neuron might supply both inner and outer hair cells has arisen. Perkins and Morest (1975), using the Golgi stain, found that the radial fibers typically innervated two inner hair cells. Occasionally, they found neurons with swellings or branches to both inner and outer hair cells in immature rats and kittens. Because branches to the inner hair cells from external spiral fibers were not found in kittens more than 1 day old, it is probable that this type of branch atrophies during maturation.

Single nerve fibers can be stained by the Golgi method in newborn guinea pigs, which are different from neonatal rats and cats in that they are mature animals. Smith and Haglan (1973), Smith (1975a), and Smith et al. (1976) stained many external spiral fibers throughout the cochlea in 1-day-old guinea pigs (Fig. 30). No branches to inner hair cells were seen. The dendritic fibers lose their myelin

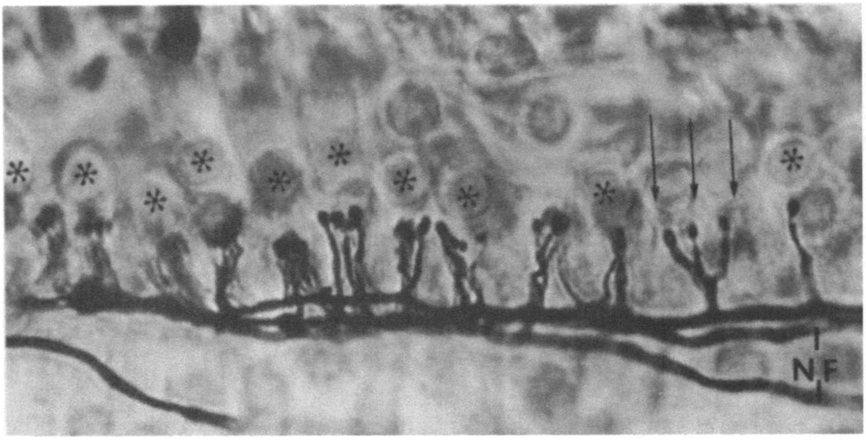

Fig. 30. Golgi-stained external spiral fibers (cochlear nerve), guinea pig, illustrating multiple nerve endings on outer hair cells from three afferent fibers. The fibers (*NF*) are separated at *right* and three nerve endings (*arrows*) from a single preterminal can be seen to end on two outer hair cells. *Asterisks* on nuclei of other outer hair cells with nerve endings. ×960

sheaths as they pierce the habenula perforata of the basilar membrane. They cross the tunnel and then turn toward the basal end, often enclosed in a Deiters cell mesaxon. The fibers in the basal turn course for some distance (up to 0.8 mm) before approaching the outer hair cells for their terminal branching. Each fiber supplies about 6 – 20 hair cells mostly in a single row. One or, less frequently, two or three nerve branches are given by a single dendrite to any one hair cell. In the apical half of the cochlea the external spiral branching pattern is somewhat different. The nerve fibers branch almost immediately after crossing the tunnel and give approximately 40 – 60 branches to two or three rows of hair cells. These multiple branches in the apex fit in with the larger number of nerve endings per outer hair cell found in the second and third rows (Smith and Sjöstrand 1961b). Recently (1979) Spoendlin has reported that the external spiral fibers are the peripheral branches of a "type II" ganglion cell, the neurons of which are unmyelinated and pseudomonopolar in the cat. He has not been able to trace the central processes of these neurons into the modiolus and believes that their axons do not project to the cochlear nucleus.

The radial nerve fibers also lose their sheaths at the habenulae and then proceed directly up to their terminations on the inner hair cells. Branching is minimal, as previously discussed. But many cochlear nerve fibers terminate on each inner hair cell. Spoendlin (1966, 1970, 1974) has shown that 90% – 95% of the cochlear nerve fibers are given to cat inner hair cells. The percentage is slightly less (85% – 90%) in the guinea pig (Morrison et al. 1975), but it seems clear that at least four-fifths of the auditory nerve dendrites supply the inner hair cells. In view of the dearth of branching, it seems apparent that a large percentage of nerve fibers is concerned with transmitting information derived from a small, restricted place on the basilar membrane, i.e., that occupied by one or two adjacent inner hair cells.

Efferent. There are two systems of efferent nerve fibers within the cochlea. The first is the efferent olivocochlear tract with terminals on the sensory cells and axodendritic synapses on the unmyelinated cochlear nerve fibers. The second is sympathetic in nature (with possibly a parasympathetic component) and has no direct relationship to cochlear duct derivatives.

The Olivocochlear Bundle. There have been innumerable morphological, physiological, and behavioral investigations on the olivocochlear tract and although we have a fairly clear idea regarding its distribution within the cochlea, we still do not understand its function. Warr (1975, 1978) has reported recently that there is a larger number of fibers in the tract, at least in some individuals, than was originally found by Rasmussen (1946). Warr (1975) utilized the retrograde transport of horseradish peroxidase to localize the neurons of origin in the brainstem. He found that more than 1500 neurons were labeled in some kittens and more than half of these were located on the side ipsilateral to the injection. In addition, many of the ipsilateral fibers came from *small* fusiform neurons located near the ipsilateral lateral superior olivary nuclei (Warr 1978).

It would be interesting to know if these brainstem numerical data and neuronal forms are the same in the chinchilla because Warr's data would fit in very well with the peripheral fiber arrangements, sizes, and individual variations in that animal. Smith and Rasmussen (1963) found that the inner spiral bundle, the tunnel bundle, and the plexus beneath the outer hair cells belonged to the olivocochlear tract of the chinchilla. Iurato et al. (1978) cut the crossed component of the tract in the chinchilla and found that most of the large vesiculated nerve endings on the outer hair cells degenerated, giving evidence that the contralateral (or crossed) tract terminated on the outer hair cells. On the other hand, the inner spiral bundle, the tunnel bundle, and a few nerve endings on outer hair cells (mostly in the apical half of the cochlea) remained in place, and it was concluded that these belonged to the ipsilateral (uncrossed) tract. The axons of the inner spiral bundle are very small (0.15 – 0.25 μm diameter) with only a few larger fibers. The number of fibers in the inner spiral bundle and tunnel bundle varies from place to place and among individuals but, in general, ranges from 250 to 350 (Iurato et al. 1978). The fiber size in the chinchilla corresponds to that of Warr's ipsilateral, small, fusiform neurons in kittens; the individual numerical variations also fit. However, it is difficult to compare numerical data from the cat and chinchilla when we have no available figures on the total number of efferent fibers entering the chinchilla's cochlea.

The Adrenergic Nerve Fibers. An adrenergic nerve plexus which coursed in the bony spiral lamina among the myelinated nerve fibers was described by Spoendlin and Lichtensteiger (1966, 1967), Terayama et al. (1966, 1968), and Spoendlin (1973). The density as well as the terminal extent of these nerve fibers appears to vary among species. Densert's (1974) observations on the rabbit's cochlea are enlightening because of the profuse adrenergic innervation of that animal's ear. He found there was a dense adrenergic nerve plexus in the spiral lamina among the radial myelinated nerve fibers and numerous fluorescent (as determined by the Falk and Hillarp method) varicosities at the habenula perforata where the cochlear nerve fibers lose their myelin sheaths. Similar findings have been reported for the cat (Densert and Flock 1974) and the guinea pig (Terayama et al.

1968). Smith (1975) has found nerve fibers (by use of the Golgi stain) in the guinea pig spiral lamina which correspond to the above in pattern and location. The fluorescent fibers in the rabbit continued peripheralward along the basilar membrane and terminated close to the vessel of the tympanic lip (below the inner hair cell) within the inner ear (Densert 1974). The adrenergic fibers in the cat do not leave the spiral lamina (Spoendlin 1973; Densert and Flock 1974). Terayama et al. (1968) have not been able to find nerve fibers adjacent to the tunnel spiral vessel in the guinea pig. The fluorescent fibers completely disappeared in the ipsilateral cochlea after unilateral cervical sympathectomy in both cat and rabbit (Densert and Flock 1974; Densert 1974).

It seems clear that there is a sympathetic nerve supply to the cochlea for which the superior cervical sympathetic trunk is the source. However, in most animals, the fibers are restricted to the modiolus and the bony spiral lamina. Some fibers follow the blood vessels; others do not. Many terminate near the habenula at the place where the nerve fibers lose their myelin sheaths. Although there is no good physiological evidence for any sympathetic effect on cochlear activity, there is morphological evidence that some type of sympathetic influence on cochlear nerve activity could occur near the habenula. Possibly the effect is such a generalized or subtle one that it has been difficult to detect by physiological investigations in the past.

The preceding descriptive review clearly shows that controversy still exists regarding important aspects of the significance and function of the sensory cells and their innervation patterns. However, some progress has been made. To summarize some important features: Most of the cochlear nerve fibers terminate on the inner hair cells and only 5% – 10% terminate on outer hair cells. The relationship of inner hair cell to nerve fiber is essentially one-to-one, so that the radial afferents collect information from a segment of the basilar membrane no greater than 30 μm in length and generally from about one-third of that (10 μm). On the other hand, outer hair cell afferents always terminate on a number of hair cells so that the external spiral afferents are collecting information from a segment of basilar membrane 50 – 300 μm in length, which furthermore may be more than 500 μm basal to the opening in the basilar membrane from which the nerve fibers emerged. Inner hair afferents terminate almost directly opposite their respective habenulae. It would appear then that inner hair cells and their nerve fibers are the frequency receptors in the cochlea, whereas outer hair cells do something else.

And, indeed, a number of functions have been proposed for the outer hair cells. The hypothesis that has greatest support at present is that the outer hair cells sensitize the inner hair cells or reduce their threshold. It is unclear exactly how this is accomplished but data from first-order neuronal responses (Dallos and Harris 1978) and behavioral audiometry (Ryan and Dallos 1975; Stebbins et al. 1979) demonstrate that the thresholds for neural responses and hearing are raised by about 50 dB when outer hair cells have been eliminated but inner hair cells and their nerve fibers remain. However, we are still left without an explanation for the function of the outer hair cell afferents. Research by Morest and Bohne (1980) indicates that the outer hair cell afferents probably do project to the cochlear nucleus. They found that small axons with a preferential projection to the dorsal cochlear nucleus degenerated after specific outer hair cell loss.

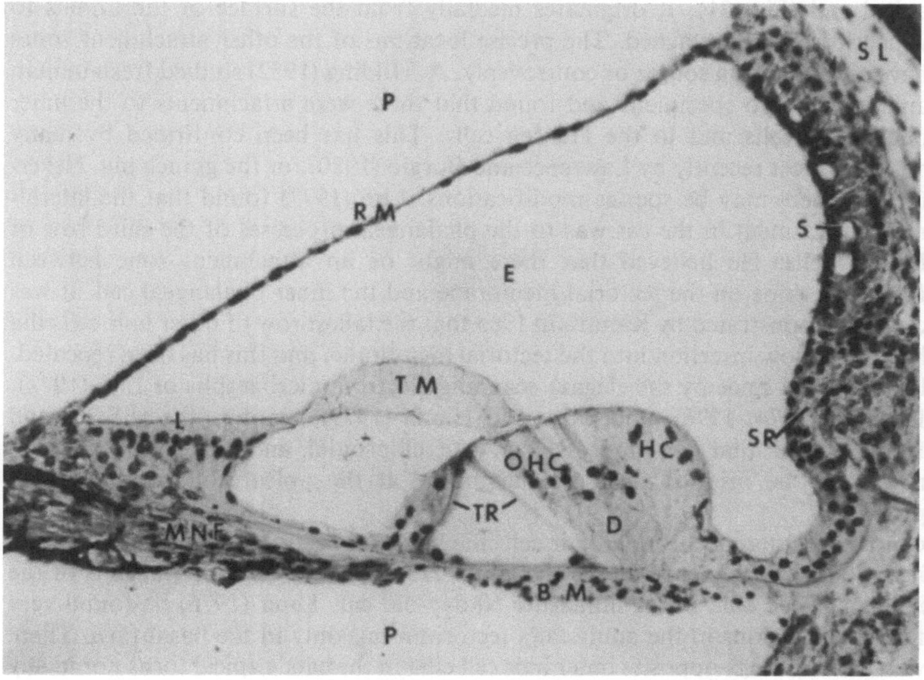

Fig. 31. Cross section of cochlear duct (light micrograph), squirrel monkey, showing relationship of the tectorial membrane (*TM*) to the limbus (*L*) and the organ of Corti. The *TM* has receded from its original attachment to the Hensen's cells (*HC*), but retains its attachment to the stereocilia of the first row of outer hair cells (*OHC*). *BM*, basilar membrane; *D*, Deiters' cells; *E*, endolymph; *MNF*, myelinated nerve fibers; *P*, perilymph; *RM*, Reissner's membrane; *S*, stria vascularis; *SL*, spiral ligament; *SR*, spiral prominence; *TR*, tunnel rods. ×335. From Smith (to be published)

5.1.4 Supporting Cells

The tunnel rod cells and the outer phalangeal cells (Deiters cells) seem to be specialized primarily for the function of skeletal support (Figs. 24, 28, 31). This is accomplished by means of bundles of cytoplasmic microtubules bound together by cross-bonds. The microtubules spread out in the reticular lamina and provide stiffness of this structure. It may be that Deiters cells have some additional function. The hypothesis has been proposed (Smith 1978) that they may be potassium regulators and thus control the ionic content in the external milieu of the outer hair cells in a manner similar to that proposed for the glial cells by Kuffler and Nichols (1976). This hypothesis has been elaborated on previously (Smith 1978) and will not be discussed further here.

5.1.5 Tectorial Membrane

The mammalian tectorial membrane is a thin, extracellular membrane which covers the top of the reticular lamina and is fairly homogeneous from basal to apical end (Fig. 31). It originates medially from the surface of the limbus to which it is firmly attached. The precise locations of the other attachment zones have always been a source of controversy. A. Hilding (1952) studied fresh human and pig embryo specimens and found that there were attachments to the inner phalangeal cells and to the Hensen cells. This has been confirmed by many authors, most recently by Lawrence and Burgio (1980) for the guinea pig. Nevertheless, there may be species modifications. Lim (1977) found that the lateralmost attachment in the cat was to the phalangeal processes of the third row of Deiters cells. He believed that there might be an attachment zone between Hensen's stripe on the tectorial membrane and the inner phalangeal cell. It was clearly demonstrated by Kimura in 1965 that the tallest row of outer hair cell cilia have a shallow insertion into the tectorial membrane, and this has been repeatedly confirmed since by the elegant scanning electron micrographs of Lim (1972), Hoshino (1974, 1976), and Saito and Hama (1979), among others. Saito and Hama found that the impressions were ellipsoidal and that even two tall rows may be inserted into the membrane at the apical coil of the guinea pig.

On the other hand, the inner hair cell cilia have not been shown to have a definite insertion into the membrane in the adult. Hoshino (1974) found imprints of the inner hair cell cilia in the immature 10-day-old cat. Later (1976) he found very faint indentations in the adult cat's tectorium but only in the basal turn. There were no markings opposite inner hair cell cilia in the cats's apical turns nor in any part of the guinea pig's cochlea. He also observed some loose fibrous strands near the inner hair cell imprints (or where they might be expected to be) which he believed could indicate that there were attachments between the inner phalangeal cells and the tectorial membrane (1976).

Lim has recently (1977) made a study of the development of the tectorial membrane and found multiple imprints of inner hair cell cilia along the outer edge of Hensen's stripe (that is, lateral to the inner hair cell zone) in the *kitten*. He suggested that they may have been formed during development when the inner hair cells shifted medialward and the cilia were dragged across the tectorial membrane, but that they had no real relation to inner hair cells in the adult.

Lim's study (1977) has been greatly informative because he compared both fresh and fixed tissue by microscopic techniques. Although he did not make measurements (which are very difficult to make on fresh material), he stated, "The detailed morphology of the fresh tectorial membrane was remarkably similar to that of the fixed TM, so much so that it was not possible to distinguish between the fixed and unfixed membranes." It is reassuring to know that the several distinguishing features of the membrane, such as the amorphous covering network on the free surface, the marginal band and net, and Hensen's stripe are not fixation artifacts. He also described a "gluey substance" which appeared late in development and helped to attach the outer hair cell cilia to the tectorial membrane. This gluey material was not found near the inner hair cell cilia. The sum of evidence seems to indicate that the inner hair cell cilia in mature animals are not as firmly inserted into the tectorial membrane as are the outer hair cells. Hoshino's observations (1976) suggest the possibility that they touch it at least in the basal coil of the cat. This difference in inner and outer hair cell cilia attachment might be an important part of the pattern which results in reduced auditory sensitivity in experimental animals whose outer hair cells have been destroyed (Ryan and Dallos 1975; Stebbins et al. 1979).

5.1.6 Basilar Membrane

The fine structure of the basilar membrane was described by Iurato (1962) as well as others, as being composed of small 100-Å filaments embedded in a ground substance. Kimura (1975) found that the small filaments were rectangular and composed of four or five subfibrils in the guinea pig. Cabezudo (1978) has confirmed this structure in the cat. It thus appears that all the extracellular fibers which are contiguous to the perilymphatic spaces may have a similar embryonic origin. This includes the extracellular fibers in the perilymphatic spaces of the vestibule (Hamilton 1967) and in the spiral ligament of the cochlea (Takahashi and Kimura 1970).

Cabezudo (1978) analyzed the basilar membrane in the cat and demonstrated that the thickness of the membrane and the relative ratio of the ground substance and fibrous elements to mesothelial cells changed from basal to apical end. The width increased gradually to almost the very tip but the total thickness (including mesothelial layer) decreased about 5 mm before the tip. Such data are important as it is possible that the latter might affect the low-frequency traveling wave.

5.2 Stria Vascularis

5.2.1 Cell Types

The stria vascularis is a well-defined group of cells which forms the lateral wall of the cochlear duct and which contains a dense intraepithelial capillary network. The tight junctions between the marginal cells at the luminal surface plus similar junctions in the basal cell layer create a third fluid compartment within the cochlea (Jahnke 1975; Reale et al. 1975), the other two being the endolymphatic and perilympathic compartments.

There are three types of cells in the stria: the marginal, the intermediate, and the basal cells (Figs. 31, 32). The marginal cells are the only ones in contact with the endolymph. Like the other cells of the epithelial cochlear duct, they are joined together by occluding junctions (Jahnke 1975; Reale et al. 1975). The marginal cells contain an abundance of endoplasmic reticulum (both smooth and rough), large mitochondria, and ribosomes suspended in a dense ground substance. The membrane profiles vary according to location. Lined vesicles (small, intermediate, and large) predominate near the surface. Golgi apparatus is found slightly deeper in the cytoplasm. Coated vesicles are scattered throughout the cytoplasm but are not very numerous. The apical plasma membrane is straight, fairly smooth, and sometimes covered by a thick surface coat. Some cells show a moderate surface pitting and this phenomenon may indicate the source of the lined vesicles. The most remarkable feature of the marginal cells is the tremendous basal expansion of the plasma membrane. Basal to the cell junctional complex, numerous long, narrow, fingerlike processes are formed. Many abut on the basal lamina of the

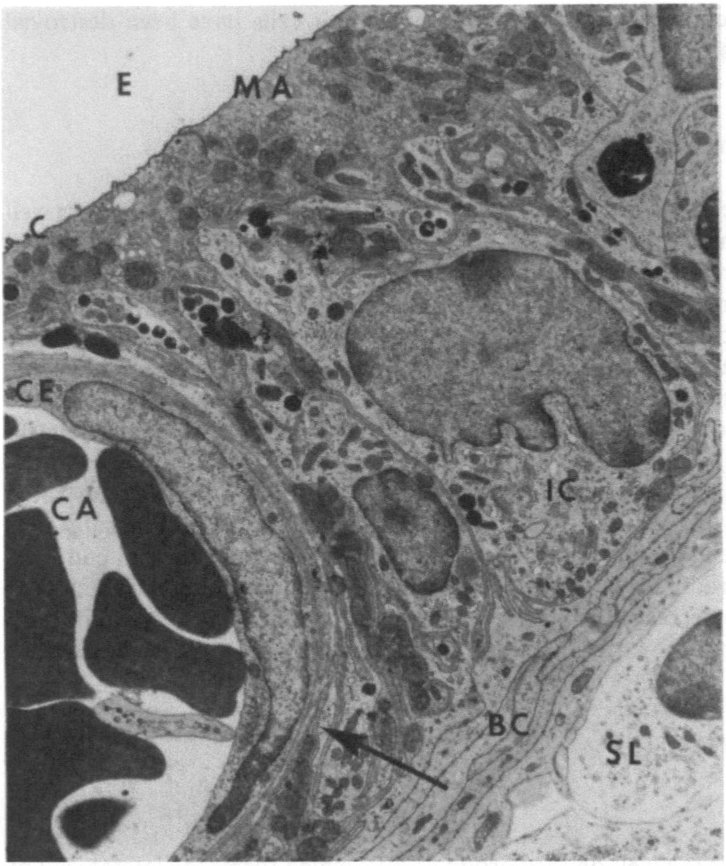

Fig. 32. TEM of stria vascularis, guinea pig. The marginal cells (*MA*) face the endolymph (*E*) and have a dense surface coating (*C*). Their basal processes (*arrow*) abut on the capillary endothelium (*CE*). The intermediate cells (*IC*) have a less dense cytoplasm and small mitochondria. The thin, flat basal cells (*BC*) are *below*. *CA*, capillary; *SL*, spiral ligament. ×6100. From Smith (to be published)

capillary endothelium; others intermingle with the shorter and less abundant processes of the intermediate cells.

The intermediate cells do not reach the endolymph surface but are located in an intermediate position between the marginal and basal cells. They do not form a continuous layer but are scattered among the marginal cells and capillaries. These cells have short fingerlike processes which also abut on the capillary endothelium. There are only a moderate number of cytoplasmic organelles in the intermediate cells. Mitochondria are much smaller than those in the marginal cells. A variable number of large vesicles containing a material slightly denser than that of the cytoplasm is usually present. The same type of vesicle, but open ended, is found at the cell membrane and it seems probable that most of these vesicles are pinocytotic vesicles. Some cells have a conspicuous amount of Golgi apparatus, as well as large membrane-bound bodies containing a dense granular material.

The basal cells are broad and flat. Some of the innermost cells in the layer send long thick columns up and between the marginal and intermediate cells but they never reach the surface. These basal cell columns separate the marginal and intermediate cells into small nests (Hinojosa and Rodriguez-Echandia 1966), but it is questionable if they ever completely isolate cell groups. The basal cells are bound together by innumerable small occluding junctions (Jahnke 1975; Reale et al. 1975) so that they form a tight, cohesive wall against the perilymph in the spiral ligament.

In the past, light microscopists differentiated between the marginal and the intermediate cells by the descriptive terms "dark cells" and "light cells," respectively, terms which are still in use. It is true that the intermediate cells are frequently paler than the marginal cells. This seems to be largely due to the difference in cytoplasmic ground substance. There is little material precipitated by the fixative in the cytoplasmic matrix of most intermediate cells. However, these terms are somewhat misleading in view of recent findings that there is considerable variation in the density of both types of cells (Pollard et al. to be published in press). These variations may be due to the respective ages of the cells, in which case some cell turnover in the stria might be postulated. Alternatively, there may be considerable variation in metabolic activity from place to place in the stria. Whatever the explanation, many intermediate cells are fully as dense as their neighboring marginal cells and can only be differentiated by position, cell form, and mitochondrial characteristics.

Melanin granules are generally present in the stria vasculares of pigmented individuals. The granules are found in the intermediate and the basal cells and it seems possible that either both or one of these cell types were derived from melanocytes during development. Neither the intermediate nor the basal cells appear to be derived from the original ectodermal otic vesicle. When blood vessels in the spiral ligament approach the lateral ectodermal wall in the process of fetal maturation, some cells from the spiral ligament also migrate inward. The migrating cells abut on the marginal cells and the capillaries and soon differentiate into the intermediate and basal cells. Typical melanocytes are rarely found in the mature cochlear duct although they are present in the modiolus and in the perilymphatic spaces of the vestibule (Smith 1956; LaFerriere et al. 1974). It seems probable that the intermediate cells in the stria vascularis originated from wandering melanocytes but the origin of the basal cells is less certain. The melanin granules are much less abundant in the basal cells and are often present in large encapsulated clusters associated with myelin figures. It is possible that the melanin has been passed from the intermediate to the basal cells, a process which commonly occurs in the skin, for example. If the latter theory is correct, then only the intermediate cells are melanocyte derivatives and the basal cells may be mesenchymal in origin. The function of the melanin is unclear. The occurrence of hypopigmentation along with deafness is a well-known phenomenon in man and animals, but it is not known if the correlation of melanin deficiency and inner ear pathology is causative or coincidental. It has been suggested that the two may reflect a common neural crest defect and not be otherwise related (Deol 1970).

It does seem fairly clear that the intermediate cells are phagocytic in nature. Duvall and Sutherland (1972) found that intravascular horseradish peroxidase passed readily across the endothelial walls of the strial capillaries and accumulated rapidly in the tissue fluid of the stria. Once in the extracellular space it was taken up by the intermediate cells.

One major activity of the stria vascularis is that of rapid ion transport which results in a high potassium concentration in the endolymph. The generation of the endocochlear potential (the high positive resting potential of the endolymph) is also a part of the process. This seems to be accomplished for the most part by the activity of the marginal cells. Recent studies after ethacrynic acid administration (Brummett et al. 1977) have shown that loss of the positive endocochlear potential was correlated with changes in cell organelles in the marginal cells, but not in the intermediate cells although all three types of cells showed a loss of cell water.

5.2.2 Capillaries

The capillary bed is an important component of the stria vascularis. It is the densest vascular network within the inner ear and it is continuous from base to apex. Arteriolar branches enter the network at frequent intervals at the vestibular edge and venous branches leave it at the tympanic edge of the structure. The resultant network forms a vascular pool which is embedded in the epithelial cells. The endothelial cells of the capillary walls have no unusual morphological characteristics. They are not fenestrated. The only fenestrated capillaries in the inner ear are found adjacent to the endolymphatic sac (Lundquist et al. 1964) and in the modiolus (Kimura and Ota 1974; Gorgas and Jahnke 1974). Nevertheless, studies by Duvall and Sutherland (1972) and Gorgas and Jahnke (1974) have shown that horseradish peroxidase (HRP) passes quickly through the strial capillaries, apparently by means of pinocytosis, rather than between the endothelial cells. On the other hand, HRP does *not* readily pass across the endothelial walls of the other capillaries in the cochlea.

5.2.3 Endocochlear Potential

It is known that the stria vascularis is actively engaged in maintenance of the high potassium concentration in the endolymph and that the high positive endocochlear potential is associated with the process. A number of excellent papers or reviews on the subject have recently appeared (Bosher and Warren 1968; Sellick and Bock 1974; Sellick and Johnstone 1975; Konishi et al. 1978; Bosher 1979). It is fairly well agreed that the transport process by which potassium is concentrated in the endolymph is responsible for the positive component of the DC potential of the endolymph. There is a negative component also which is apparent with anoxia or administration of certain drugs (Bosher et al. 1973; Brummett et al. 1977). There is some controversy regarding the source of the negative potential. Bosher (1979) has recently produced evidence that it is probably a diffusion potential. It has also been proposed that it may be associated with the outer hair cells of the organ of Corti (Honrubia et al. 1976).

It is interesting that the *high* positive DC potential has been found only in mammalian endolymph. Schmidt and Fernandez (1962) measured a positive potential from the cochlear endolymph of several reptilian species but it was generally low. The potential ranged from +2 to +7 mV among turtles, lizards, snakes, and crocodiles. Recent studies by Peterson et al. (1978) have shown that endolymph potassium from the alligator lizard's cochlea exceeds 100 mM/liter in some individuals. The identity of the cells which are responsible for the potassium concentration as well as for the positive DC potential in reptiles is unknown as no specialized cells such as found in the stria vascularis of mammals have been described for these animals.

The endocochlear potential of the bird's ear (Schmidt and Fernandez 1962) is slightly higher. It averages +9 mV with a high value of +16. There are no recent values for potassium in avian endolymph but it is probable that the potassium concentration is higher than in perilymph. The avian cochlear duct contains an extensive cell mass, the tegmentum vasculosum, which forms one wall of the duct, and it has been assumed for years that it is similar in function to the mammalian stria vascularis. This assumption is based on the fact that the tegmentum vasculosum includes two types of cells, one with dense cytoplasm, large mitochondria, and profuse basal infoldings, and a second type with a paucity of cell organelles. These two types are morphologically similar to the mammalian marginal and intermediate cells, respectively. However, there is no intraepithelial vascular bed in the bird. The capillaries are located in the connective tissue, comparable to the situation in the vestibule and the spiral prominence. Apparently an important feature that has evolved in the mammalian cochlea is the capillary network which is embedded in the marginal and intermediate cells and sealed off from the perilymph by the basal cells. This could permit a rapid fluid and ion interchange between capillary and cells. No such structural relationship is present in lower animals.

5.3 Spiral Prominence

The spiral prominence comprises a small group of cells located between the stria vascularis and the external spiral sulcus (Figs. 24, 33). These cells are frequently disregarded in discussions of the endocochlear potentials or of maintenance of the chemical composition of endolymph. Yet they are similar in structure to the cells with basal infoldings in the vestibule which are generally considered to be "secreting" cells there. Furthermore, the spiral prominence cells have a separate capillary system of their own. The structure was named a "prominence" because the cells bulge into the endolymph and form a spiral protrusion. The bulge is caused by the capillary which is located just beneath the basal lamina of the epithelium and which forms a spiral vessel or network of capillaries (Smith 1951). This vessel is not necessarily continuous from base to apex, anymore than are the capillaries beneath the basilar membrane, but it is a separate entity not connected directly to either the network of the stria vascularis or the arteriovenous arcades in the spiral ligament.

The cells of the spiral prominence are cuboidal in shape and have some microvilli on the surface facing the endolymph. Their apical ends are joined together by occluding junctions as are the other epithelial cells of the membranous labyrinth. Below the junctional complex the cell membrane forms a number of thick

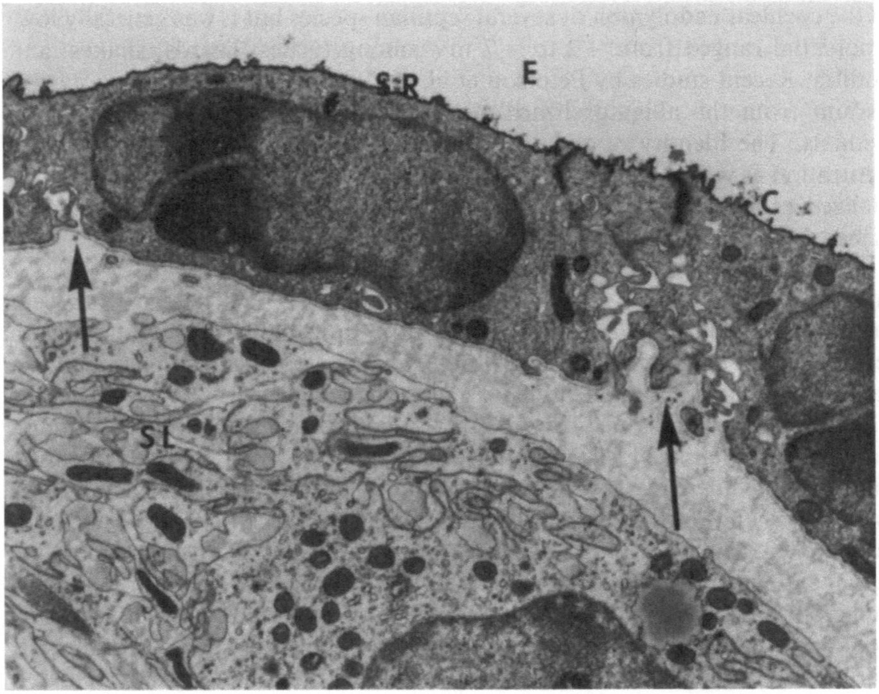

Fig. 33. TEM of spiral prominence, guinea pig. The epithelial cells (*SR*) have lateral projections (*arrows*) and a dense cytoplasm. *C*, surface coat; *E*, endolymph; *SL*, spiral ligament. ×10300

thumblike projections. These basal projections are shorter and not nearly as extensive as those formed by the marginal cells of the stria but they apparently fulfill the same purpose, i.e., increasing the area of the lower cell surface. In addition, the processes of these cells project into large, irregular fluid spaces. The processes are more numerous on the sides of the cells than on the basal surface. The largest part of the basal surface is fairly flat and abuts on a basal lamina. The cells of the spiral ligament are often at some distance from the prominence epithelial layer, separated from it by a space loosely filled with filaments and perilymph.

The cytoplasm of the cells is dense and contains many membranes in the form of vesicles, cisternae, and Golgi apparatus. The endoplasmic reticulum is both smooth and rough. In addition, free ribosomes and microfilaments are present within the cytoplasm.

It is likely that the stria vascularis and the spiral prominence act in concert or are synergistic. There is evidence that ferritin is taken up by the spiral prominence cells from the endolymph as well as from the perilymph (Hinojosa 1972). Konishi et al. (1978) found that potassium passed readily from scala tympani into the endolymph and hypothesized that it passed across the stria. It seems more logical to assume that it passes or is transported across the spiral prominence cells. It is proposed that the cells of the spiral prominence act in a manner similar to that of the morphologically comparable cells of the vestibule (Smith 1970) and that they are active in the transport of potassium ions into the cochlear endolymph. If this is correct the spiral prominence may make a small contribution to the positive endocochlear potential. The endovestibular potential is approximately +2 to +4 (Smith et al. 1958; Kusakari and Thalmann 1976). It has been recently hypothesized by Marcus and Thalmann (1980), on the basis of Konishi et al.'s (1978) data, that there is an additional potassium pump, besides the one in the stria vascularis, in the cochlea but no cell source was suggested. The best candidate for the proposed pump would be the cells in the spiral prominence.

6 Concluding Remarks

One important feature that is present in all auditory mechanoreceptors discussed in this review is a stiffness in the apical part of the structure. In the single ciliated insect receptors this stiffness is accomplished by means of microtubules in the neuronal receptor cell. The reptilian and avian auditory receptor organs are compact structures, and the sensory cells are supported greatly by the mass of the structure. In these animals the microtubules are located in the supporting cells and only scattered tubules are to be found in the receptor cells. Packed microfilaments are present in the stereocilia and cuticular plates. However, it is in the mammalian organ of Corti that we can really appreciate the property of skeletal support that has long been hypothesized as an important function of the microtubules. There are large fluid spaces between the sensory cells within the organ of Corti, and the structure would undoubtedly collapse without the microtubular bundles within the tunnel rod cells and the Deiters cells. In addition, the extension of the tubular bundles into the surface of the organ (the reticular lamina) provides an added apical stiffness.

The remarkable variability in the form and attachments of the lizard tectorial membranes emphasizes its importance. The larger reptilian papillae (the crocodilia) have tectorial membranes which cover all the hair cells and closely resemble those of birds. Our present data show that birds adopted a tectorium of rather consistent structure: a membrane that is thick an held firmly in place by the supporting cell rings which encircle each hair cell. In addition, the tallest cilia on each hair cell are firmly inserted into the tectorial membrane. Actually, a tectorium with multiple, firm attachments seems to be the only practical way to maintain contacts with the many hair cells massed together on the avian basilar membrane. The mammalian tectorial membrane has attachments to the supporting cells at the inner and outer spiral edges of the organ of Corti but no accessory attachment to the reticular lamina. The tangential arrangement of the tectorial membrane filaments across the surface of the organ of Corti makes a "bias tape" pattern around the outer edges of each coil. This arrangement coupled with the limbic attachment at the central pole of the modiolus may help to hold it in place. It is interesting at this point to look back down the evolutionary scale to reptiles and find that lizards and mammals have an amazing similarity. Some hair cell cilia are firmly attached to the tectorium (outer hair cells) whereas others (inner hair cells) are not.

One outstanding feature of the mammalian cochlea is its remarkable structural consistency from lower to higher mammals, from small to large cochleae. There are always four rows of hair cells (one inner and three outer) on any one segment of basilar membrane from basal to apical end. From mouse to man the structure of the organ of Corti varies little.

There are a number of other comparisons one might make, but it is hoped that this short review will be stimulating enough that the readers themselves will ask the questions and try to find the answers.

References

Baird IL (1970) The anatomy of the reptilian ear. In: Gans C, Parsons T (eds) Biology of the reptilia, vol 2, Academic Press, New York, pp 193 – 275

Baird IL (1974) Anatomical features of the inner ear in submammalian vertebrates. In: Keidel WD, Neff WD (eds) Anatomy, physiology (ear). Springer, Berlin Heidelberg New York (Handbook of sensory physiology, vol 5/1, pp 159 – 212)

Bekesy G (1960) Experiments in hearing. McGraw-Hill, New York, pp 504 – 506

Boord RL (1964) The number and diameter of myelinated fibers in the statoacoustic nerve of the adult pigeon. Am Zool 4:101

Bosher SK (1979) The nature of the negative endocochlear potentials produced by anoxia and ethacrynic acid in the rat and guinea pig. J Physiol 293:329 – 345

Bosher SK, Warren RL (1968) Observations on the electrochemistry of the cochlear endolymph of the rat: a quantitative study of its electrical potential and ionic composition as determined by means of flame spectrophotometry. Proc R Soc Lond [Biol] 171:227 – 247

Bosher SK, Smith C, Warren RL (1973) The effects of ethacrynic acid upon the cochlear endolymph and stria vascularis. Acta Otolaryngol 75:184 – 191

Bredberg G (1968) Cellular pattern and nerve supply of the human organ of Corti. Acta Otolaryngol [Supp] (Stockholm) 236:1 – 135

Brummett R, Smith CA, Ueno Y, Cameron S, Richter R (1977) The delayed effects of ethacrynic acid on the stria vascularis of the guinea pig. Acta Otolaryngol 83:98 – 112

Cabezudo L (1978) The ultrastructure of the basilar membrane in the cat. Acta Otolaryngol 86:160 – 175

Dallos P, Harris D (1978) Properties of auditory nerve response in absence of outer hair cells. J Neurophysiol 41:365 – 383

Densert O (1974) Adrenergic innervation in the rabbit cochlea. Acta Otolaryngol 78:345 – 356

Densert O, Flock Å (1974) An electron microscopic study of adrenergic innervation in the cochlea. Acta Otolaryngol 77:185 – 197

Deol MS (1970) The relationship between abnormalities of pigmentation and of the inner ear. Proc R Soc Lond [Biol] 175:201 – 217

Dooling RJ (1973) Behavioral audiometry with the parakeet (*Melopsittacus undulatus*). J Acoust Soc Am 53:1757 – 1758

Dooling RJ, Saunders JC (1975) Hearing in the parakeet (*Melopsittacus undulatus*): Absolute thresholds, critical ratios, frequency difference limens and vocalizations. J Comp Physiol Psych 88:1 – 20

Dooling RJ, Mulligan JA, Miller JD (1971) Auditory sensitivity and song spectrum of the common canary (*Serinus canarius*). J Acoust Soc Am 50:700 – 709

Dooling RJ, Zoloth SR, Haylis JR (1978) Auditory sensitivity, equal loudness, temporal

resolving power and vocalization in the house finch (*Carpodacus mexicanus*). J Comp Physiol Psych 92:867 – 876
Dunn RA, Morest DK (1975) Receptor synapses without synaptic ribbons in the cochlea of the cat. Proc Natl Acad Sci 72:3599 – 3603
Duvall AJ, Sutherland CR (1972) Cochlear transport of horseradish peroxidase. Ann Otol Rhinol Laryngol 81:705 – 714
Flock Å (1964) Structure of the maculae utriculi with special reference to directional interplay of sensory responses as revealed by morphological polarization. J Cell Biol 22:413 – 431
Flock Å (1965) Electron microscopic and electrophysiological studies on the lateral line canal organ. Acta Otolaryngol [Suppl] 199:1 – 90
Flock Å (1971) Sensory transduction in hair cells. In: Loewenstein WR (ed) Principles of receptor physiology. Springer, Berlin Heidelberg New York (Handbook of sensory physiology, vol 1, pp 396 – 441)
Flock Å, Wersäll J (1962) A study of the orientation of the sensory hairs of the receptor cells in the lateral line organ of fish, with special reference to the function of the receptors. J Cell Biol 15:19 – 27
Ghiradella H (1971) The fine structure of the Noctuid moth ear. I. The transducer area and connections to the tympanic membrane in *Feltia subgothica haworth*. J Morphol 134:21 – 45
Goldberg JM, Fernandez C (1971) Physiology of peripheral neurons innervating semicircular canals of the squirrel monkey. I. Resting discharge and response to constant angular acceleration. J Neurophysiol 34:635 – 684
Goodley LB, Boord RL (1966) Quantitative analysis of the hair cells of the auditory papilla of the pigeon. Am Zool 6:542
Gorgas K, Jahnke K (1974) The permeability of blood vessels in the guinea pig cochlea. II. Vessels in the spiral ligament and the stria vascularis. Anat Embryol [Berlin] 146:33 – 42
Gray EG (1960) The fine structure of the insect ear. Philos Trans R Soc Lond [Biol] 243:75 – 94
Gulley RL, Reese TS (1977) Regional specialization of the hair cell plasmalemma in the organ of Corti. Anat Rec 189:109 – 124
Hamilton DW (1967) Perilymphatic fibrocytes in the vestibule of the inner ear. Anat Rec 157:627 – 640
Hamilton DW (1969) The cilium on mammalian vestibular hair cells. Anat Rec 164:253 – 258
Held H (1926) Die Cochlea der Säuger und der Vögel, ihre Entwicklung und ihr Bau. In: Bethe A, von Bergmann G, Embden G, Ellinger A (eds) Handbuch der Normalen und Pathologischen Physiologie mit Berücksichtigung der Experimentellen Pharmakologie, Springer, Berlin, pp 476 – 526
Hilding A (1962) Studies on the otic labyrinth. 1. On the origin and insertion of the tectorial membrane. Ann Otol Rhinol Laryngol 61:354 – 370
Hinojosa R (1972) Electron microscope studies of the stria vascularis and spiral ligament after ferritin injection. Acta Otolaryngol 74:1 – 14
Hinojosa R, Rodriguez-Echandia ER (1966) The fine structure of the stria vascularis of the cat inner ear. Am J Anat 118:631 – 664
Holton T, Weiss T (1978) Two tone rate suppression in lizard cochlear nerve fibers, relation to receptor organ morphology. Brain Res 159:219 – 222
Honrubia V, Strelioff D, Sitko S (1976) Physiological basis of cochlear transduction and sensitivity. Ann Otol Rhinol Laryngol 85:697 – 710
Hoshino T (1974) Relationship of the tectorial membrane to the organ of Corti. A scanning electron microscope study of cats and guinea pigs. Arch Histol Jpn 37:25 – 39
Hoshino T (1976) Attachment of the inner sensory cell hairs to the tectorial membrane. A scanning electron microscope study. Otol Rhinol Laryngol 38:11 – 18

Iurato S (1962) Functional implications of the nature and submicroscopic structure of the tectorial and basilar membranes. J Acoust Soc Am 34:1386–1395
Iurato S (1967) III. Light microscopic features. In: Iurato S (ed) Submicroscopic structure of the inner ear. Pergamon, New York, pp 18–38
Iurato S, Smith CA, Eldredge DH, Henderson D, Carr C, Ueno Y, Cameron S, Richter R (1978) Distribution of the crossed olivochochlear bundle in the chinchilla's cochlea. J Comp Neurol 182:57–76
Jahnke K (1975) The fine structure of freeze-fractured intercellular junctions in the guinea pig inner ear. Acta Otolaryngol [Suppl] 336:1–40
Katsuki J, Suga N (1960) Neural mechanism of hearing in insects. J Exper Biol 37:279–290
Kikuchi K, Hilding DA (1965) The development of the organ of Corti in the mouse. Acta Otolaryngol 60:207–222
Kimura RS (1965) Hairs of the cochlear sensory cells and their attachment to the tectorial membrane. Acta Otolaryngol 61:55–72
Kimuara RS (1975) The ultrastructure of the organ of Corti. Int Rev Cytol 42:173–222
Kimura RS, Ota CY (1974) Ultrastructure of the cochlear blood vessels. Acta Otolaryngol 77:231–250
Kimura RS, Ota CY, Takahaski T (1979) Nerve fiber synapses on spiral ganglion cells in the human cochlea. Ann Otol Rhinol Laryngol [Suppl] 62:1–17
Kimura RS, Schuknecht HF, Sando I (1964) Fine morphology of the sensory cells in the organ of Corti of man. Acta Otolaryngol 58:390–408
Konishi M (1970) Comparative neurophysiological studies of hearing and vocalizations in song birds. Z Vergl Physiol 66:257–272
Konishi M (1973) How the owl tracks its prey. Am Scientist 61:414–424
Konishi T, Hamrick PE, Walsh PG (1978) Ion transport in guinea pig cochlea. I. Potassium and sodium transport. Acta Otolaryngol 86:22–34
Kuffler SW, Nichols JG (1976) From neuron to brain. A cellular approach to the function of the nervous system. Sinauer, Sunderland, Mass.
Kusakari J, Thalmann R (1976) Effects of anoxia and ethacrynic acid upon ampullar endolymphatic potential and upon high energy phosphates in ampullar wall. Laryngoscope 86:132–147
Lawrence M, Burgio PA (1980) Attachment of the tectorial membrane revealed by scanning electron microscope. Ann Otol Rhinol Laryngol 89:325–330
LaFerriere KA, Arenberg IK, Hawkins JE, Johnson L-G (1974) Melanocytes of the vestibular labyrinth and their relationships to the microvasculature. Ann Otol Rhinol Laryngol 83:685–695
Liberman MC (1978) Auditory-nerve response from cats raised in a low noise chamber. J Acoust Soc Am 63:442–455
Liberman MC (1979) Terminal portions of afferent fibers to inner hair cells in the cat as studied in serial ultrathin sections. Soc Neurosci Abs 5:25
Liberman MC (1980) Morphological differences among radial afferent fibers in the cat cochlea: An electron microscopic study of serial sections. Hearing Res 3:45–63
Lim DJ (1972) Fine morphology of the tectorial membrane. Its relationship to the organ of Corti. Arch Otolaryngol 96:119–215
Lim DJ (1977) Fine morphology of the tectorial membrane. INSERM 68:47–60
Lorente de No R (1937) The sensory endings in the cochlea. Laryngoscope 47:373–377
Lowenstein O, Wersäll J (1959) A functional interpretation of the electron microscopic structure of the sensory hairs in the cristae of the elasmobranch, *Raja clavata,* in terms of directional sensitivity. Nature 184:1807–1808
Lundquist PG, Kimura R, Wersäll J (1964) Ultrastructural organization of the epithelial lining in the endolymphatic duct and sac in the guinea pig. Acta Otolaryngol 57:65–80

Manley GA (1970) Frequency sensitivity of auditory neurons in the caimen cochlear nucleus. Z Vergl Physiol 66:251 – 256
Marcus DC, Thalmann R (1980) Comments concerning a possible independent potassium pump in the cochlear duct. Hear Res 2:163 – 165
Michelsen A (1974) Hearing in invertebrates. In: Keidel WD, Neff WD (eds) Anatomy physiology (ear). Springer, Berlin Heidelberg New York (Handbook of sensory physiology, vol 5/1, pp 389 – 422)
Miller MR (1966) The cochlear duct of lizards. Proc Calif Acad Sci 33:255 – 359
Miller MR (1973a) Scanning electron microscope studies of some lizard basilar papillae. Am J Anat 138:301 – 330
Miller MR (1973b) A scanning electron microscope study of the papilla basilaris of *Gekko gecko*. Z Zellforsch 136:307 – 328
Moran DT, Chapman KM, Ellis RA (1971) Fine structure of the cockroach campaniform sensilla. J Cell Biol 48:155 – 173
Morest DK, Bohne B (1980) Differential degeneration in the brain of chinchillas following inner and outer hair cell damage by cochlear stimulation [Abstr] Assoc Res Otolaryngol Jan:11
Morrison D, Schindler RA, Wersäll J (1975) A quantitative analysis of the afferent innervation of the organ of Corti in guinea pigs. Acta Otolaryngol 79:11 – 23
Mulroy MJ (1974a) Cochlear anatomy of the alligator lizard. Brain Behav Evol 10:69 – 87
Mulroy MJ (1974b) Intracellular electric responses to sound in a vertebrate cochlea. Nature 249:482 – 485

Perkins RE, Morest DK (1975) A study of cochlear innervation patterns in cats and rats with the Golgi method and Nomarski optics. J Comp Neur 163:129 – 158
Peterson SK, Frishkopf LS, Lechene C, Oman CM, Weiss TF (1978) Element composition of inner ear lymphs in cats, lizards and skates determined by electron probe microanalysis of liquid samples. J Comp Physiol 126:1 – 14
Pollard T, Smith C, Brummett R (to be published) The effects of low dose ethacrynic acid on the guinea pig cochlea with special reference to variations in the normal stria vascularis. Acta Otolaryngol

Rasmussen GL (1946) The olivary peduncle and other fiber projections of the superior olivary complex. J Comp Neur 84:141 – 220
Reale E, Luciano L, Franke K, Pannese E, Wermbter G, Iurato S (1975) Intercellular junctions in the vascular stria and spiral ligament. J Ultrastruct Res 53:284 – 297
Reese TS (1965) Olfactory cilia in the frog. J Cell Biol 25:209 – 230
Retzius G (1884) Das Gehörorgan der Wirbeltiere. II. Das Gehörorgan der Reptilien, der Vögel und der Säugetiere. Samson and Wallin, Stockholm
Roeder K, Treat AE (1961) The detection and evasion of bats by moths. Am Sci 49:135 – 148
Rosenbluth J (1967) Cochlear ganglion. In: Iurato S (ed) Submicroscopic structure of the inner ear, Pergamon, New York, pp 239 – 245
Ryan A, Dallos P (1975) Effect of absence of cochlear outer hair cells on behavior auditory threshold. Nature 253:44 – 46

Sachs MB, Young ED, Lewis RH (1974) Discharge patterns of single fibers in the pigeon auditory nerve. Brain Res 70:431 – 447
Saito K, Hama K (1979) Scanning electron microscopic observations of the under surface of the tectorial membrane. J Electron-Microsc (Tokyo) 28:36 – 42
Schmidt RS, Fernandez C (1962) Labyrinthine DC potentials in representative vertebrates. J Cell Phys 59:311 – 322
Schwartzkopff J (1968) Structure and function of the ear and of the auditory brain areas in birds. In: de Reuck AVS, Knight J (eds) Hearing mechanisms in vertebrates. Little, Brown, Boston, pp 41 – 58

Sellick PM, Bock GR (1974) Evidence for an electrogenic potassium pump as the origin of the positive component of the endocochlear potential. Pfluegers Arch 352:351–361
Sellick PM, Johnstone BM (1975) Production and role of inner ear fluid. Prog Neurobiol 5:337–362
Sjöstrand FS (1956) The ultrastructure of cells as revealed by the electron microscope. Int Rev Cytol 5:455–533
Smith CA (1951) Capillary areas of the cochlea in the guinea pig. Laryngoscope 61:1073–1095
Smith CA (1956) Microscopic structure of the utricle. Ann Otol Rhinol Laryngol 61:450–470
Smith CA (1961) Innervation pattern of the cochlea: The internal hair cell. Ann Otol Rhinol Laryngol 70:504–527
Smith CA (1970) Extrasensory cells of the vestibule. In: Paparella M (ed) Biochemical Mechanisms in hearing and deafness, Thomas, Springfield, pp 171–185
Smith CA (1973) The efferent neural supply to the vertebrate ear. Adv Otorhinolaryngol 20:296–310
Smith CA (1975a) Innervation of the cochlea of the guinea pig by use of the Golgi Stain. Ann Otol Rhinol Laryngol 84:443–459
Smith CA (1975b) The inner ear: Its embryological development and microstructure. In: Tower DB (ed) The nervous system, vol 3, Raven, New York, pp 1–18
Smith CA (1978) Structure of the cochlear duct. In: Naunton RE, Fernandez C (eds) Evoked electrical activity in the auditory nervous system. Academic Press, New York, pp 3–19
Smith CA (to be published) Structure of the cochlear duct. In: Beagley H (ed) Audiology and audiological medicine. Oxford Press, Oxford
Smith CA, Davis H, Deatherage BH, Gessert CF (1958) DC potentials of the membranous labyrinth. Am J Physiol 193:203–206
Smith CA, Dempsey EW (1957) Electron microscopy of the organ of Corti. Am J Anat 100:337–368
Smith CA, Haglan BJ (1973) Golgi stains on the guinea pig organ of Corti. Acta Otolaryngol 75:203–210
Smith CA, Rasmussen GL (1963) Recent observations on the olivocochlear bundle. Ann Otol Rhinol Laryngol 72:489–507
Smith CA, Sjöstrand FS (1961a) A synaptic structure in the hair cells of the guinea pig cochlea. J Ultrastruct Res 5:184–192
Smith CA, Sjöstrand FS (1961b) Structure of the nerve endings of the guinea pig cochlea as studied by serial sections. J Ultrastruct Res 5:523–556
Smith CA, Tanaka K (1981) Further observations on the avian basilar papilla. Abs. Assoc Res Otolaryngol
Smith CA, Takasaka T (1971) Auditory receptor organs of reptiles, birds and mammals. In: Neff W (ed) Contributions to sensory physiology. Academic Press, New York
Smith CA, Cameron S, Richter R (1976) Cochlear innervation: Current status and new findings by use of the Golgi stain. In: Hirsh SK, Eldredge DH, Hirsh, IJ, Silverman SR (eds) Hearing and Davis, Essays honoring Hallowell Davis. Washington University Press, St. Louis
Spoendlin H (1966) The organization of the cochlear receptor. Adv Otorhinolaryngol 13
Spoendlin H (1970) Structural basis of peripheral frequency analysis. In: Plomp R, Smoorenburg GF (eds) Frequency analysis and periodicity detection in hearing. Sijthoff, Leiden, pp 2–36
Spoendlin H (1973) Autonomic nerve supply to the inner ear. In: deLorenzo AJ (ed) Vascular disorders and hearing defects. University Park Press, Baltimore, pp 93–109
Spoendlin H (1974) Neuroanatomy of the cochlea. In: Zwicker E, Terhardt E (eds) Facts and models of hearing. Springer, New York, pp 18–32

Spoendlin H (1979) Neural connections of the outer hair cell system. Acta Otolaryngol 87:381 – 387

Spoendlin H, Lichtensteiger W (1966) The adrenergic innervation of the labyrinth. Acta Otolaryngol 61:423 – 434

Spoendlin H, Lichtensteiger W (1967) The sympathetic nerve supply to the inner ear. Arch Klin Exp Ohr Nas Kehl Kopfheilkd 189:346 – 359

Stebbins W (1970) Studies of hearing and hearing loss in the monkey. In: Stebbins W (ed) Animal psychophysics. Appleton-Century-Crofts, New York, p 153

Stebbins WC, Hawkins JE, Johnson L, Moody DB (1979) Hearing thresholds with outer and inner hair cell loss. Am J Otolaryngol 1:15 – 27

Suga N, Campbell HW (1967) Frequency sensitivity of single auditory neurons in the gecko *Coleonyx variegatus.* Science 157:88 – 90

Takahashi T, Kimura RS (1970) The ultrastructure of the spiral ligament in the rhesus monkey. Acta Otolaryngol 69:46 – 60

Takasaka T, Smith CA (1971) The structure and innervation of the pigeon's basilar papilla. J Ultrastruct Res 35:20 – 65

Tanaka K, Smith CA (1975) Structure of the avian tectorial membrane. Ann Otol Rhinol Laryngol 84:287 – 297

Tanaka K, Smith CA (1978) Structure of the chicken's inner ear: SEM and TEM study. Am J Anat 153:251 – 272

Terayama Y, Holz E, Beck C (1966) Adrenergic innervation of the cochlea. Ann Otol Rhinol Laryngol 65:69 – 86

Terayama Y, Yamamoto K, Sakomoto T (1968) Electron microscopic observations on the postganglionic sympathetic fibers in the guinea pig cochlea. Ann Otol Rhinol Laryngol 77:1152 – 1170

Warr BW (1975) Olivocochlear and vestibular efferent neurons of the feline brain stem: Their location, morphology and number determined by retrograde axonal transport and acetylcholinesterase histochemistry. J Comp Neurol 161:159 – 182

Warr BW (1978) The olivocochlear bundle: It's origins and terminations in the cat. In: Naunton RF, Fernandez C (eds) Evoked electrical activity in the auditory nervous system. Academic Press, New York, pp 43 – 65

Weiss TF, Mulroy MJ, Altmann DW (1974) Intracellular responses to acoustic clicks in the inner ear of the alligator lizard. J Acoust Soc Am 55:606 – 619

Weiss TF, Mulroy MJ, Turner RG, Pike C (1976) Tuning of single fibers in the cochlear nerve of the alligator lizard: Relation to receptor morphology. Brain Res 115:71 – 90

Weiss TF, Peake WT, Ling A, Holton T (1978) Which structures determine frequency selectivity and tonotopic organization of vertebrate nerve fibers? Evidence from the alligator lizard. In: Naunton RF, Fernandez C (eds) Evoked electrical activity in the auditory nervous system. Academic Press, New York, pp 91 – 112

Wever EG (1967a) The tectorial membrane of the lizard ear: Types of structure. J Morphol 122:307 – 320

Wever EG (1967b) The tectorial membrane of the lizard ear: Species variations. J Morphol 123:355 – 372

Wever EG (1978) The reptile ear: Its structure and function. Princeton University Press, Princeton, New Jersey

Young D (1973) Fine structure of the sensory cilium of an insect auditory receptor. J Neurocytol 2:47 – 58

Progress in Sensory Physiology

Editors:
H. Autrum,
D. Ottoson,
E. R. Perl,
R. F. Schmidt

Editor-in-Chief:
D. Ottoson

The objective of this series is to provide concise and critical information on current advances in the different domains of sensory physiology. It will be of interest to all who want to keep abreast of the latest developments in the field – from the level of the receptor to that of the cortex, including neuropsychologic and psychophysical aspects.

Particular emphasis will be placed on progress in the following areas:

- Transducer functions and membrane properties of receptors
- Synaptic mechanisms in transmission of sensory information
- Information processing at higher levels of the nervous system
- Functional and morphologic organization of sensory systems
- Motor mechanisms and sense-organ functions
- Psychophysical and neuropsychologic aspects of sensory functions
- Behavior and sensory functions
- Comparative aspects of the structure and function of sense organs.

Volume 1

With contributions by numerous experts

1981. 72 figures, 6 tables. VII, 179 pages
ISBN 3-540-08413-4

Contents:

Springer-Verlag
Berlin
Heidelberg
New York

Handbook of Sensory Physiology

Editorial Board:
H. Autrum,
R. Jung,
W. R. Loewenstein,
D. M. MacKay, H. L. Teuber

Volume 1
Principles of Receptor Physiology
Editor: W. R. Loewenstein
1971. 262 figures. XII, 600 pages
ISBN 3-540-05144-9

Volume 5 (In 3 Parts)
Auditory System
Part 1:
Anatomy, Physiology (Ear)
Editors: W. D. Keidel, W. D. Neff
1974. 305 figures. VIII, 736 pages
ISBN 3-540-06676-4

Part 2:
Physiology (CNS), Behavioral Studies, Psychoacoustics
Editors: W. D. Keidel, W. D. Neff
1975. 209 figures, 12 tables. VII, 526 pages
ISBN 3-540-07000-1

Part 3:
Clinical and Special Topics
Editors: W. D. Keidel, W. D. Neff
1976. 343 figures. VII, 811 pages
ISBN 3-540-07129-6

Volume 6:
Vestibular System
Part 1
Basic Mechanisms
Editor: H. H. Kornhuber
1974. 251 figures. VIII, 676 pages
ISBN 3-540-06889-9

Part 2:
Psychophysics, Applied Aspects and General Interpretations
Editor: H. H. Kornhuber
1974. 198 figures. VIII, 680 pages
ISBN 3-540-06864-3

Volume 9:
Development of Sensory Systems
Editor: M. Jacobson
1978. 231 figures, 9 tables. IX, 469 pages
ISBN 3-540-08632-3

Springer-Verlag
Berlin
Heidelberg
New York

MIX
Papier aus verantwortungsvollen Quellen
Paper from responsible sources
FSC® C105338

If you have any concerns about our products,
you can contact us on
ProductSafety@springernature.com

In case Publisher is established outside the EU,
the EU authorized representative is:
Springer Nature Customer Service Center GmbH
Europaplatz 3, 69115 Heidelberg, Germany

Printed by Libri Plureos GmbH
in Hamburg, Germany